Laying Down Arms to Heal the Creation-Evolution Divide

For Grace,

Gary

Laying Down Arms to Heal the Creation-Evolution Divide

Gary N. Fugle

Foreword by
Darrel R. Falk

WIPF & STOCK · Eugene, Oregon

Wipf and Stock
An Imprint of Wipf and Stock Publishers
199 W. 8th Ave., Suite 3
Eugene, OR 97401

www.wipfandstock.com

Front cover photographs and design by Jacob Reid Design.

ISBN 13: 978-1-62564-978-2

Manufactured in the U.S.A.   12/10/2014

Unless otherwise indicated, Bible quotations are from the HOLY BIBLE. NEW INTERNATIONAL VERSION. Copyright © 1973, 1978, 1984 by International Bible Society. Published by The Zondervan Corporation, Grand Rapids, MI.

Contents

List of Figures

List of Tables

Foreword

THAT GARY FUGLE WOULD choose the title, "Laying Down Arms" for a book about biology and the Christian faith will, for some, cause a certain amount of unease right from the start. Fugle, after all, believes in evolution. In Ephesians 6 does not Paul call on Christians to "take up the full armor of God" in order to be protected from a world subject to the force of darkness? Belief in evolution is frequently considered to be the root of that darkness in today's world. To many Christians, the concept of evolution seems to be among the most effective ideological tools that Satan has at his disposal. If we were to lay down our armor on this particular front, such people believe, we would be opening ourselves up to the demise of Christianity itself. The Bible, as the authoritative Word of God, would be the first to go, followed by human uniqueness, a meaningful basis for morality and ethics, belief in miracles, acceptance of the resurrection, and the concept of an eternal human soul. "Laying Down Arms" then, in many minds, would be tantamount to no longer protecting oneself against that which threatens to destroy the purpose and meaning of their lives and even more importantly, the lives of their children and grandchildren.

This book exists to show that this concern, although fully understandable, is unnecessary and likely even counter-productive. In a patient, respectful, and loving manner, Fugle shows why none of the precepts that lie at the heart of a Christ-centered life and the theology on which it is based are changed or challenged if God chose to create through the evolutionary mechanism. Building upon his career as an ecologist, evolutionary biologist, and an effective science communicator, as well as his life centered in Christian community, Fugle presents the case for laying down arms in the battle against evolution. He shows why it is not only safe, but right to do so. He presents his position with a unique knack for making biological concepts accessible to a general audience, and a profound understanding of the biblical and theological issues involved. His love for biology is contagious, his winsome affection for the Church, an inspiration.

His call for Christian commitment to truth is clear and persistent. He writes with an assurance that *both* the scientific method and the Bible point us to a deeper understanding of the Creator's activity in this world. Each informs the other—science tells us *how* because it is a gift from God; it is a reliable tool for coming to understand God's world. The Bible tells us *who*, also because it is a gift from God; it reveals the reality of the living Word through whom, in whom, and for whom all things were created. Just as Christians are called to study God's Word to explore the wonder-filled ramifications of *who* God is, Fugle rightly implies that they are called to open their minds to what God says through the tools of science by which we are able to explore the wonder-filled ramifications of *how* God's world has been and is being created.

Still, even after exploring the matter in the way described above, Christians will not all come to see things in the same way, nor as I see it, should they. This book is not a call to accept evolutionary creation as one's own view; it is a call to lay down arms. It is important that we all come to understand that there is not as much at stake as many have thought. There need not be a battle that pits the theory of evolution against a biblically based understanding of the Christian faith. And there certainly ought not be any sort of war over the issue within the Christian community. So even though we discuss our differences as each of us seeks fuller understanding, the conversation must take place bathed in a spirit of oneness, immersed in sacred purpose. This book then, is a diligent attempt to help us all be truth-seekers, while never overlooking the call to be people who are known by our love. In my view, it accomplishes both purposes very well.

This book is not only for Christians. There are many individuals who, based on the biological evidence for evolution conclude that biblically-based Christianity is now obsolete. This view is philosophically, biblically, and theologically naïve and it behooves those who hold it to do some exploring of their own. The lines that too often have been drawn between biology and the biblically-grounded Christian faith do not exist. There is no battle to be waged between faith and science. They are friends not foes. Fugle clearly demonstrates that the two are compatible. Certain overzealous skeptics continually lob science-fueled grenades at Christians, but what Fugle does so well in contrast is to show that mainstream science enriches Christian worship and strengthens Christian belief. Swords are turned into plowshares and spears into pruning hooks.

When battles are over, typically one side is declared the victor and the other side is forced to disarm. Fugle shows that this battle is different. He shows that here there is a peace accord in which both sides win. Mainstream biology is provided with a purpose-filled basis for the amazing story that it is able to tell. Christian faith is provided with a deeper understanding of what it means to speak of God as Creator. With that the arms can be laid down and an era of Christ-centered, peace-filled co-existence can begin.

Darrel R. Falk

Former President of the BioLogos Foundation.
Professor Emeritus of Biology, Point Loma Nazarene University.

Acknowledgements

THIS BOOK PROJECT BEGAN many years ago when I first sensed there might be a reason that God placed me in the worlds of both evolutionary biology and evangelical Christianity. A large number of individuals have positively influenced its development since that inception. There is really no way I could mention all who have cheered me on or offered something significant (even if I could remember everyone). So, I begin here with wide appreciation for all the family, friends, students and colleagues who have contributed to making this book what it is. I have been blessed and I know it!

Thanks to the scientists at Butte College who shared my passion for the natural world and embraced and valued my dual citizenship as a professional biology teacher and Christian, especially Paul Mason, Katherine Newman and Albin Bills. Katya Yarosevich has been a special help as a fellow traveler in science and Christian faith; I thank her for being with me through celebrations and struggles and always encouraging me. I thank David Danielson for how deeply he has understood my intent with the book and for his unbridled enthusiasm and support for it.

Many evenings of lively discussion were spent with my friends in the Chico Triad on Philosophy, Science and Theology. All the participants helped mature and refine my thinking and spurred me on with consistent acceptance and backing. Thanks especially to Greg Cootsona, Ric Machuga and Dan Barnett for being the solid core of the group, as well as reviewing earlier portions of my manuscript, providing concrete help when I needed it, and for fully embracing me. I benefited from the brilliant young mind of Michael Fitzpatrick, who read the entire book and gave many helpful suggestions. Michael also inspired me by always expressing confidence in my abilities and the worth of this project.

I would never have begun or completed this endeavor without the full life I've experienced with my extended church family, especially those at Ridge Presbyterian Church in Paradise, CA. Thank you to those who have walked with me for all these years, especially Chris and Teresa Reid, Craig and Toni Bolger, and Doug and Becky Wion. To Chris Reid, I still marvel at how God made us incredibly alike in many ways and yet different in others, so that we have powerfully commiserated and celebrated together. Of course Chris read a version of the manuscript and provided uplifting encouragement.

Darrel Falk has been a particular help in bringing this project to completion. His positive response to my work rallied me at a time when I was not sure about its future and I am so thankful for the time he has invested to support me and the book. Darrel's gentle, patient and humble approach to all things, and his unwavering communication for love and unity within the church, have inspired me.

Thank you to Jacob Reid for the front cover design. The imagery is both stunning and calming, and provokes a wonderful sense of future resolution. Jake's varied artwork displays a unique giftedness that continues to surprise me.

Finally, I thank my family. My son and daughters, Jedidiah, Hannah and Mary, have made sacrifices; I deeply know this and I am grateful. In the end, a book is a very small thing compared to the joy, fullness, and meaning that they are to me. My wife, Brenda, is the greatest blessing of my life. Certainly, whatever good there is in this book, it emanates from her as well as me. She has been vital as an editor and sounding board. Most important, she has always been my strongest supporter and steadfastly believed in the vision for this book.

PART I

Introduction

THE BATTLES HAVE GONE on for as long as anyone can remember. Vocal leaders from both sides rally their troops with warnings about the critical concerns at stake and the injustices committed by the opposition. But a different voice is being heard of hope and peace and celebration. It reminds us that deep Christian faith and progressive science were long championed together as noble goals to be pursued with passion and excitement, that the current conflicts are not the way it has always been. It proclaims an uplifting and deeply relevant message that biblical Christianity and evolutionary biology may be brought together in a vibrant and meaningful way without compromising the core convictions of either endeavor. If given good reasons to do so, are we willing to lay down our arms to end the wars and affirm a positive future?

1

The Journey

A GRADUATE STUDENT IMMERSED in evolutionary biology at a major university was on the verge of completing his doctoral degree when Jesus captured his heart and he turned over his life to the sovereign God of the universe. Christians rejoice as they picture another person joining the family of God, while non-Christians wonder what might have led to such an unexpected conversion. But what did this person do with his mechanistic view of life and all those years investigating biological evolution? This is my story.

I grew up in southern California. I attended a Presbyterian church for several years during early elementary school until I persuaded my mother I'd rather stay home with my dad on Sunday mornings than go to church with her. I learned a few things about Christian faith, but never had much commitment to it. My preteen years are best characterized by my playing hard with my two brothers and our friends at whatever sport happened to be in season. I was also engaged in an especially active Boy Scout troop of over 50 kids and this gave me a vivid and enjoyable exposure to natural environments. In high school, I was a better-than-average student, but I cared more about my social life and my basketball and tennis teams than my education.

My horizons expanded when I attended college classes and became inspired by a variety of talented instructors. I developed a love of learning and a deep fascination for the natural world. Almost any biology subject intrigued me. I spent endless hours studying the colorful collage of animals and plants scattered across our planet to understand how they were put together and how they worked. At the top of my list were courses and research that provided insight into what I observed in the outdoors where there are wonderful intertwined relationships among living things. I found particular satisfaction in the evolutionary concepts that pervaded many of my classes as I saw they had the power to explain so much about the complexities of nature. My learning experiences were mixed with a lot of camping and backpacking, especially through the Sierra Nevada mountain range.

My enthusiasm in these years led to academic success, professional research publications, and a bachelor's degree from the University of California, Santa Barbara. The latter was capped by the Biological Sciences Department's "Most Outstanding Graduating Senior" award. As I entered graduate school, my highest ideal was the pursuit of knowledge through study and research (how exciting to push beyond the boundary of what we currently know and understand!), and my greatest love and inspiration were the beauty and grandeur in nature (my heart soared to see the brilliant colors of a Western Tanager in the morning sun or to take in the expansive view of a wooded valley that lay below a peak I had just climbed).

Then, as my graduate studies progressed at the University of California, Davis, and back at UC, Santa Barbara, my love and excitement for biology continued but my worldview started to change. My idealistic sense of our capacity to know and understand was slowly deflated by my observations of an all-too-frail and fallible (read "human") nature found in even the greatest minds within my discipline. I became increasingly disenchanted with my mechanistic view of life and its inability to encompass all that I had experienced. My heart softened to a larger realm of possibilities. In my early years of college I was far from neutral toward Christians; I rejected them as weak, anti-intellectuals who stood in direct opposition to my way of thought. But when the timing was right, the message of Jesus and a God that is bigger than all I know and understand became oddly appealing.

True to my training, I dove into research about Jesus, the Bible, and the specific beliefs of Christian faith. I read through the Bible and all kinds of investigative books, had endless discussions with friends, attended Bible studies and seminars, and I watched and wondered. I found it easy and surprisingly comforting to acknowledge the reality of my own shortcomings[1] and I was drawn by the truth and promise in the message of new life and a new worldview offered through confession and faith in Jesus.[2] I recognized that to believe in Christ could not merely be an intellectual ascent while my life continued along pretty much as usual. Instead, to trust in Jesus would be to turn my world upside-down, where nothing else would matter by

1. Rom 7:15–25, Luke 5:27–32.
2. Rom 3:21–24, 8:5–6; John 15:5–8.

comparison (as one old hymn expresses it, "Thou my best thought by day or by night . . . ") and my life's calling would be to serve and depend upon the Lord of the universe.[3] I took that step of faith in October 1980.

I went on to complete my Ph.D. and published additional professional articles. I discovered during graduate school that I experienced the greatest joy and fulfillment sharing my love for biology in teaching, rather than in the details of research. I have endeavored to impart my enthusiasm ever since as a professor in the California community college system, developing and teaching courses across a spectrum from human anatomy and physiology to detailed surveys in zoology and botany. I have been fortunate and honored to receive different educational awards, such as "Faculty Member of the Year" and "Educator of the Year" from Butte College in northern California. I have been especially grateful that my students have given me top ratings over these many semesters. I still have that passion for biology and the natural world.

I have also been an excited participant in conservative Christian churches for over 30 years.[4] I have immersed myself in studying the Bible and I have developed a deep appreciation and understanding of a God-centered, biblical worldview. I have led home fellowships and Bible studies and taken major roles in various other ministries (men's retreats, pastor searches, etc.). I regularly lead worship services at my church as a speaker and musician and consider it a particular joy to construct meaningful music experiences in these settings. Most important, I have lived and worshipped alongside fellow Christians for all these years.[5]

The Conflict

It would have been impossible at the time of my Christian conversion to be unaware of the ongoing conflict between proponents of creationism and advocates of evolutionary thought. The debate has raged for years and continues today. Although there are various approaches to the creation-evolution topic, including formulations of "old-earth creationism" and "intelligent design," the most vocal and well-publicized positions have always been two extreme views: (1) Christians who hold to special creation

---

3. Matt 4:10,11:28–30; Col 3:23–25.

4. I was baptized as a member of a Southern Baptist church, later participated in different non-denominational Bible churches (including a new church plant), and, most recently, I've spent many years in a congregation of the conservative Presbyterian Church in America (PCA).

5. Acts 2:42, Col 3:12–16, Heb 10:24.

events accomplished by God over six 24-hour days several thousands of years ago, and (2) scientists who view our existence and everything else in the universe, including biological evolution, as the product of exclusively mechanistic processes operating over billons of years. Given the state of my understanding in the early 1980s, I was faced with the question of how I or anyone else could possibly exist simultaneously in these two worlds.

I have considered what the Bible says about the origin of the universe and life itself. I am guided by the understanding that the words of this book are God-inspired[6] and that one cannot pick and choose what to believe or the integrity of the whole document becomes suspect. The acts of creation described in Genesis 1 and 2 are well-known and are an entrenched part of Western cultural awareness. I will discuss the specifics of these accounts later but there is little debate they communicate that God is the sovereign, powerful Creator of all things. This message is not limited to the early chapters of Genesis; it penetrates both the Old and New Testaments. The Bible repeatedly identifies God as the Creator[7] and specifically states that he made all aspects of this world and the heavens,[8] including humans.[9] It further indicates that God's action was not only a past endeavor, but that his involvement continues and is essential today.[10] In the book of Acts (17:24–25), the apostle Paul explains to the intellectuals of Athens:

> The God who made the world and everything in it is the Lord of heaven and earth and does not live in temples built by hands. And he is not served by human hands, as if he needed anything, because he himself gives all men life and breath and everything else.

There can be no doubt that the Bible exclaims God as the creative and sustaining force behind our existence. But what exactly does Scripture say about *how* God created? Are evolutionary ideas excluded?

I have considered whether God would have me abandon my understanding of evolution and much of my biological background. I have evaluated the tenets of evolution and the evidence for it in the natural world. I've pondered evolution's predictive power and the beauty with which it ties together biological knowledge. Is it all a misconstrued fabrication? Could such a major idea in science be false? I also considered all the professional biologists within my discipline at colleges and universities throughout our

6. Matt 5:17–18, 2 Tim 3:16, 2 Pet 1:20–21.

7. Gen 14:19, Deut 32:16, Isa 40:28, Rom 1:25, 1 Pet 4:19.

8. Exod 20:11; Ps 95:5, 104:24, 136:7; Isa 45:12,18; Jer 27:5; John 1:2; Col 1:16; Rev 10:6.

9. Gen 9:6, Ps 100:3; 139:13.

10. Acts 17:28, Col 1:17, Heb 1:3.

country (and the world). Could they be misled by an entrenched tradition or blinded by atheistic leanings? I conclude, with confidence, that there is enormous validity to evolutionary thought. It is an intellectually rigorous scientific construct and provides a meaningful avenue for making sense of the natural world. Evolutionary biologists are, for the most part, honest scientists without an inherent hidden agenda (although a large number of them overstep what can be extrapolated from what we know). I am convinced that God would not have me discard my understanding of evolution.

I have come to a very simple conclusion: both creation and evolution are true! I am guided by the premise that there will never be conflict between biblical accounts and scientific findings of biology. I believe the Bible and the natural world have the same divine author and both will reveal that author's character.[11] This means that Christians are always invited to celebrate exciting new truths uncovered by ongoing biological research. It also means that scientists are free to marvel at and be inspired by the God of the Bible and his hand in nature. I obviously reject a mechanistic view of life devoid of God, as well as any religious position that denies the discoveries of modern scientific investigation.

Simply stated, my position is that God, as the sovereign Creator of all things, utilized evolutionary processes to develop the diversity of life on earth. I uphold a solid biblical doctrine of Creation and its significant implications regarding God's control and relationship to his creation, including humans. I simultaneously understand that there have been complex evolutionary connections among organisms over very long periods of time. I perceive that God certainly created the physical parameters and processes of the natural world that have led to the evolution of various forms, but I also accept that his guiding hand has directed the course of these mechanisms. I will present further details of my reasoning in the chapters that follow.

This book is about evolutionary ideas and how they relate to Christian faith. My path has brought me to a relatively unusual position in which I simultaneously participate in the scientific community of evolutionary biology *and* the religious community of conservative evangelical Christians in America. It is from this vantage point that I hope to communicate to a spectrum of readers. My primary audience is Christians, most of whom are relatively uninformed about the specifics of evolution and have not heard good reasoning for how it can be accepted along with a Christian worldview. I write as a fellow believer and I aim to convey a scientific understanding of evolution that can be appreciated with excitement rather than apprehension.

11. Ps 19:1–2, Rom 1:20 and see chapter 4.

At the same time, my desired audience also includes non-Christians who already accept evolutionary principles. For these individuals, my goals are to communicate a meaningful understanding of evangelical Christian thought, to explain what is at stake for believers in creation-evolution discussions, as well as to present how a personal Christian faith can be embraced along with evolutionary ideas. I acknowledge upfront that this book has a prominent Christian tone and focus with reference to common views within the evangelical church, the use of Bible quotations and specific appeals to Christian believers. But this is not meant to exclude non-Christians. Rather, I believe it is the most effective way to communicate key issues to both Christians and non-Christians. Ultimately, I offer encouragements and challenges to both audiences.

I address this book especially to anyone who senses that the real truth about creation and evolution lies in a balanced view that includes valuable aspects of both concepts. This includes the significant number of scientists and others who already hold various intellectual positions that allow Christianity and evolution to come together and who welcome additional food for thought. (The existence of this latter group is generally not well-publicized, but it is larger than most people imagine).

I do not intend this book for staunch proponents of six-day, young-earth creationism who hold unswervingly to their position that God created all things over six literal days several thousand years ago. In the same vein, I do not have committed atheists in mind. I am not interested in an adversarial debate as to who is right or wrong about creation and evolution among people who are convinced that these are two mutually exclusive positions. This kind of conflict has characterized public discussion for far too long. It has, in my opinion, produced little of value and has instead deepened the chasm that exists between two extremes.

In direct contrast, I am enthusiastically interested in a dialogue among individuals who are softened to the possibility of reconciliation in which the powerful message of Christian faith and the fascinating scientific understanding of evolution are integrated together. I have been greatly encouraged by frequent encounters with Christians who are eager for a more balanced perspective to become influential in the Church, as well as by regular interactions with non-believers who are excited to hear that evolution does not have to conflict with Christian beliefs. So, if you are interested and willing, I invite you on this journey to see how creation and evolution are in harmony.

2

Why Should You Care?

WITH SUCH APPARENTLY DISPARATE views between biblical creationism and evolutionary theory, it might seem easiest for people to simply choose one or the other. As straightforward as this may sound, it is an incomplete assessment of the debate where choices have greater repercussions than most people recognize. I'm sure that readers of this book start from a variety of places. Some are eager for the dialogue to begin. Others are not so sure and bring doubts, a sense of concern, or even very strong reservations. Ultimately, I believe there are important reasons for almost everyone to be drawn into the discussion.

The Range of Beliefs

In my many years with the creation-evolution issue, I have found that every adult I've ever asked has heard of the controversy and formulated at least some opinions about it. Why is that? It isn't, of course, a common attraction to evolutionary biology. Instead, it's because the topic touches on many of the deepest and most fundamental questions about our human existence. Why and how are we here? Do we have a purpose in life? What is our relationship to the natural world in which we live? Is there a God and, if so, what is this God like? If evolution implies that our human existence and the diversity of life on earth is the result of undirected mechanical processes, then the answers to these questions are strikingly different than explanations derived from spiritual or religious perspectives. Everyone seems to understand, either explicitly or subconsciously, that this is what is at in the heart of the creation-evolution debate. It is no surprise that people care deeply and have strong opinions about such things. But the real answers to these questions have never been based on making a "God versus evolution" decision.

The Gallup Organization has assessed the beliefs of Americans pertaining to creation and evolution on 11 different occasions between 1982 and 2012.[1] The polls have consistently shown that only up to 16% of Americans (varying from 9 to 16% over these years) accept a purely mechanistic view that humans and other living forms developed over millions of years with no involvement from God. Another contrasting group of up to 47% (ranging from 40 to 47%) believe that "God created human beings pretty much in their present form within the last 10,000 years." Where is the representation of all the other Americans? Up to 40% (varying from 35 to 40%) responded to the third choice in the surveys: "Human beings have developed over millions of years from less advanced forms of life, but God guided this process." A different 1997 Gallup poll showed that, even among practicing scientists, 40% claimed the position that God directed evolution.[2]

These results reveal that, despite what is commonly depicted, many individuals recognize that a spiritual reality of a sovereign God can be combined with scientific understandings of evolutionary change. This means the general view I am offering in this book is not entirely new, radical, or unique (although I certainly hope to add new and clarifying perspectives). There are a variety of reasons why the position is not more recognized. These include: (1) media that generally prefer to promote controversy between polar beliefs, (2) young-earth creationist organizations that have had enormous influence on Christians by pushing the message that evolution is equivalent to a godless philosophy, and (3) scientists who have brazenly communicated that evolutionary discoveries demonstrate there is no God, just as conservative creationists have warned. However, it is apparent from the poll results that it is possible to be freed from a false dichotomy dictating that one must choose between a spiritual view of life and modern scientific discoveries.

It is important for many Christians to know that the majority of those in the Gallup polls who accept that God has directed evolution also identify themselves as professing Christians, and this includes many evangelicals. In some Christian circles, in fact, the notion of compatibility between biblical beliefs and biological evolution is a foregone conclusion and discussions focus on the details of how faith and evolutionary science most appropriately interrelate. For example, Robert Russell, director of the influential

1. Newport, "Creationist View."
2. Witham, "Many Scientists."

interdisciplinary Center for Theology and The Natural Sciences, has written, "Today the majority of scholars who take seriously the mutually constructive interaction between theology and science support the core conviction that evolution is God's way of creating life."[3]

The fact that so many Americans embrace the notion that God directed evolution may also be of significance to nonreligious individuals. People are often surprised at how few Americans actually take a fully atheistic view. The truth is that the vast majority of individuals seek out spiritual realities and that many religious perspectives can be compatible with evolutionary theory. Most students and scientists with whom I have spoken are very aware of the association of young-earth creationism with Christian faith and they will flatly reject Christianity on this basis. On the other hand, many find an appeal in the approach offered in this book that communicates the joys of Christianity while also allowing an open-minded view of evolutionary processes. A key goal for me in writing this book is to communicate the deep significance and beauty that I experience in both my Christian faith and in my study of biology.

Surely Not Conservative Evangelical Christians!

Readers may be thinking that this is all fine and good for a general discussion, but that there is no way committed conservative Christians who believe in an inerrant Bible are going to accept modern evolutionary biology. Contrary to that assumption, I have lived alongside evangelical believers for quite a long time and know that many are quite open to hear how biblical faith and evolution can be compatible. Large numbers of conservative Christians are not fully comfortable with strict young-earth creationist formulations communicated by peers and from the pulpit. Even though they would mark the Gallup poll selection, "God created human beings pretty much in their present form within the last 10,000 years" (especially when the question pertains directly to the emotionally charged origin of *humans*), they are not ready to reject the evolutionary communications put forth by modern scientists. I direct my presentation in this book particularly to this evangelical culture and the sensitivities of Christians who have not been given a clear opportunity to understand and accept evolutionary biology.

It is fairly common for Christians to simultaneously hold creationist and evolutionary understandings without a clear framework for how the two fit together. Some will hear inspiring messages in church about creation events or Adam and Eve and then later enjoy televised nature programs with

3. Russell, "Special Providence," 336.

strong evolutionary themes. Others happily participate as members of conservative creationist-oriented church families while also working alongside valued professional colleagues who embrace all aspects of modern science. For many, these contrasting worldviews are allowed to coexist in unresolved tension precisely because it seems like they should fit together, even though it is unclear how they might be harmonized. This compartmentalization of ideas is not at all surprising since Christian churches typically do not provide a reasoned discussion of how scientific discoveries (especially evolutionary discoveries) may be integrated with Christian beliefs. Further, the scientific community is certainly not equipped or expected to explain how evolution fits with deeply held faith. Many Christians, including conservative evangelicals, would welcome an accessible dialogue to help them move on from this dilemma.

A corollary and very significant group of Christians are those who have had their faith challenged by modern scientific information. A common scenario is a young adult with strong creationist convictions who is introduced to evolutionary theory and/or Big Bang cosmology in college classes. It is not unusual for such students to be profoundly persuaded by scientific evidences for these ideas and the reasonableness of the professors who communicate them. This can be an extremely traumatic experience, especially if the only "advice" the individual can get is to choose between staying true to the Bible and the faith of his or her church family or accepting very convincing scientific communications about the world in which we live. As a college biology professor, I have seen this conflict over and over again. What a terrible choice for an individual to face! Other individuals encounter this same crisis of faith as they enter careers in technology or science or sometime later in their life as they hear accumulating information about an ancient universe and the relatedness of all living things. Is this really an either-or choice between belief in God or assent to the modern discoveries of science? I welcome anyone who has been in this place of struggle and longs for discussion about how such things can be brought together.

Thoughts for All Christians

Even if some Christians are motivated to proceed, many others have no proclivity at all to investigate evolution. At first blush, I acknowledge that there may seem to be little incentive to fit evolution into a creationist worldview already developed from particular readings of the Bible. For believers, the Bible and scientific information like evolution do not have equivalent value. The authority of Scripture and devotion to Jesus are a way of life upon which

Christian believers depend on a daily basis. The Bible is God's message to his people. Evolutionary ideas, in comparison, have comparatively little bearing on everyday life, even for avid advocates. If pressed with a truly irresolvable conflict, a Christian will fall on the authority of Scripture over anything the scientific community proposes. Nonbelievers often do not understand the nature of this commitment. Many Christian church attendees have heard young-earth, six-day creation accounts for as long as they can remember in Sunday school and from the pulpit. It is part of what they "know" and there seems to be little reason to consider alternative views. Other Christians allow that the earth may be very old or that some evolutionary ideas seem reasonable, but their key understanding is that all things are here by God's design. By comparison, the details of evolutionary theory have far less value or significance and correspondingly it would seem there is little reason to study or understand it.

In addition, what many, if not most, Christians have heard about evolution is that it is not at all supported by natural evidence, and that those who promote it are not to be trusted. I have been around creation-evolution confrontations a very long time and I know these points are key understandings for many Christian. Creationist media repeatedly communicates that there are serious challenges to evolutionary theory to which scientists have no response. We hear that there is significant debate among biologists about previously held evolutionary positions. Given that the mainstream scientific community promotes biological evolution as a major unifying principle of great importance, many Christians conclude there must be a hidden bias that is suppressing dissension. This presumed bias has been variously connected to anti-church sentiments, atheist leanings among scientists, or even frightening human applications of "survival of the fittest."

Christians with these understandings enter into a broad American culture that has widely accepted that life on earth is united through a long evolutionary history. This acceptance is especially present in academic institutions, across the secular media, and in the public eye. What sort of response should a Christian believer have when the topic of evolution arises? Many Christians strongly uphold their understanding of biblical creationism to counter their perceived bias of scientists and to demonstrate faith in and devotion to Jesus and the Bible. In many respects, this response is understandable. But is it the right thing to do? Is it honorable to reject evolutionary ideas on these bases? Although this approach is motivated by good intentions, I must suggest it is primarily misguided and ultimately harmful.

First, let me urge Christian readers to consider that there is a more reasonable alternative picture of the mainstream scientific community than that painted by the majority of creationist communications. The real reason so many scientists embrace evolutionary theory is simply that they have been persuaded by an overwhelming collection of legitimate natural evidence. Scientists are drawn to the great value of evolutionary concepts because these approaches have the power to explain what we see in the natural world. Mainstream scientists are generally trustworthy, open-minded professionals, the ranks of which include large numbers of Christians as well as non-Christians. From a broad view of who these scientists are (and I am one of them), it is an extreme charge to insist that such an enormous group of people across a very wide discipline has deceived the rest of society. In contrast, the number of scientists who actually question the major tenets of evolutionary theory is very small.

Similarly, I suggest that the voices of six-day, young-earth creationists and intelligent design advocates have not been widely suppressed or ignored by mainstream scientists; rather, they have been evaluated and deemed incomparable and incompatible with the scientific validity and value of evolutionary theory. This is a significant distinction. I acknowledge and will later discuss very real, unacceptable atheistic tendencies within the scientific community, and I will offer challenges to change this status quo. However, these atheistic inclinations do not negate the scientific conclusions of a broad evolutionary past. Although young-earth and intelligent design creationists continue their criticisms of evolutionary biology, the vast majority of scientists know there are numerous written refutations to their challenges. I will elaborate on these points in later chapters.

My simple but deeply important suggestion is that, in contrast to claims of a lively debate and widespread disagreement about the validity of evolution among scientists, evolutionary theory remains a profoundly useful, foundational principle in biology for legitimate reasons. One of the changes that will bring healing and an end to the creation-evolution wars is an understanding within the Christian church that most scientists are simply pursuing their professions and are not the enemy of biblical Christian faith.

The Impact of Christians

Further, and especially important for believers, the character of a Christian's response to biological evolution should be significantly molded by considering one's effect on the world. Certainly, a primary aspiration for the Christian church is effectiveness at communicating the gospel of Jesus and inviting others to share in the life in which we find fullness and meaning. The apostle Peter exhorts, "But in your hearts set apart Christ as Lord. Always be prepared to give an answer to everyone who asks you to give the reason for the hope that you have" (1 Pet 3:15). In this light, how important is any particular creationist position? The central communication of Christianity is Jesus. Is something like a belief in six literal days of creation or an insistence on intelligent design presentations in schools part of the gospel message upon which a person's faith depends? Indeed, they are not. Is a limitation on how much organisms can evolve important? If there is legitimate evidence for evolutionary ideas, as I have suggested, then it follows that Christians should speak very cautiously and openly on the subject (even if emotions pull in a different direction), rather than deter someone from the focus of our faith.

The Bible clearly declares God the Creator of all things and there are very important theological implications that follow from this declaration, but nowhere does Scripture call us to defend a particular viewpoint on the specifics of creation events. Some creationists argue that the validity of the Bible and foundational Christian theology collapse without a six-day, young-earth interpretation of Genesis 1. I will fully address various aspects of this position in later chapters. In short, conservative Christian scholars have never agreed on a young-earth declaration. In fact, the majority Christian perspective throughout history has been one of tolerance and uncertainty on how one should read the biblical references to God's creative actions. As an example of historical focus, the Apostles Creed, a very early, widely accepted summation of Christian beliefs, includes only "I believe in God the Father Almighty, maker of heaven and earth . . . " There is no mention of how God made things or when and how long it took to accomplish any action (even though the Creed's authors had full knowledge of the first chapter of Genesis). There is great value in acknowledging that the topic of *how* God created is outside the central Christian message.

In some evangelical Christian communities, a six-day, young-earth creationist position has been elevated to the point that it is viewed as an essential element of the gospel or an unalterable fundamental belief. A significant number of evangelical churches and bible colleges include it in their foundational doctrinal statements. This has had the unfortunate effect of driving away individuals who might otherwise consider Christian faith. In fact, vocal young-earth creationists are a significant reason the church has lost the respect and presence it once had in the scientific world and in many intellectual circles. I can personally attest to the alienation of numerous colleagues and friends. They recognize that many Christians take an immovable creationist view by reliance upon a specific interpretation of the Bible and will not consider an open dialogue concerning evolutionary evidences. Why then should they consider a discussion about the merits of Jesus? Is this where God would have his church today? Should Christians just stick with simple creation accounts and indirectly (or directly) abandon those with evolutionary views as unfit for the message of Jesus?

Even as Christians argue for less-strict creationist positions that accommodate an old age for the earth or reason for an intelligent design approach in science classrooms, they should be cautious in the very same fashion as to how their message is conveyed. Although a Christian will always proclaim God as the sovereign creator and sustainer of all things, the specifics of creation events and how God might be integrated into science should be discussed with much less certainty.

I am drawn to passages in the New Testament that indicate how Christians can stray from the central message that we are saved by faith in Jesus. The apostle Paul explained to the churches in Corinth and Galatia that the act of circumcision was not an essential condition for a Christian.[4] Many insisted that this traditional Jewish practice was required of believers but Paul argued this was an unnecessary burden that was not needed for Christ's acceptance. "For in Christ Jesus, neither circumcision nor uncircumcision has any value. The only thing that counts is faith expressing itself through love" (Gal 5:6). I will argue that to demand a specific view of how God created is much like demanding a specific position on circumcision; both are off-track from the essence of Christ. Consider also the Pharisees in Jesus' time who demanded proper adherence to the laws they had devised out of a zealousness for God, but Jesus pointed out they were missing the heart of the Scriptures. Is the zeal of some creationists, particularly young-earth creationists, missing the point in a comparable manner and doing more harm than good? I have been around the creation-evolution debate long enough to know that some will cry that I am a "wolf in sheep's clothing" and question my faith in Jesus (it still hurts after all these years). I can only ask that people follow a popular phrase of past years and consider in their hearts, "What would Jesus do?"

I appeal to fellow Christians to adopt open-minded and tentative positions regarding creation and evolution. Be ready "to give the reason for the hope that you have," but exercise caution before expressing certainty in any particular version of creationism. I would never expect Christians to compromise what would be against God's essential message, but we should be ready to meet people where they are on issues that do not challenge the central integrity of that message. Other Christian pastors, theologians, and scientists have pleaded for humility and appealed for believers to be sensitive to the validity of scientific discoveries in order to build a bridge

---

4. 1 Cor 7:17–19, Gal 5:1–6.

to non-Christians. This appeal has only partially penetrated the organized influence of creationist groups on the evangelical church.

I understand it may be difficult for many Christians to reconsider long-held creationist views to accommodate evolutionary ideas (for some, it's easier to just think God did it by instantaneous special creation), but I encourage believers to more warmly accept people who do embrace evolution. It is the job of the church to reach out to these individuals, not the job of the latter to penetrate through a creationist barrier to get to Jesus. If for no other reason than this, Christians should be determined to actively consider whether creation and evolution might be reconciled.

Natural Science, Faith, and Scientists

I can envision that my scientific colleagues will appreciate how I have encouraged skeptical Christians to reconsider their views on evolutionary biology. I am hoping most of these colleagues will also understand that the general scientific community has some parallel attitudes and approaches that can be adjusted. Allow me to flip the discussion around to communicate reasons why non-Christians, and all scientists in particular, should be motivated toward constructive dialogues of evolutionary and creationist perspectives.

I fully appreciate that science is a wonderful avenue for the discovery of truth. As a student and practitioner of biology, I have long enjoyed the quest to uncover the working secrets of the natural world. I have been especially intrigued by an understanding that the physical parameters and processes of nature are out there operating whether we as humans are looking to learn about them or not. It is a huge and exciting playground of discovery! Biologists are given an unparalleled opportunity to conduct research to unravel the choreography of biological mechanisms, from the remarkable intricacies of genetic coding on DNA molecules, to the elaborate workings of a human kidney, to the delicate, unfolding flower parts on a brilliant red rose and much, much more. I have found rich joy and satisfaction in detailed explanations for the form and operation of living things.

Science is especially appealing on an intellectual level as a useful methodology and way of knowing. We focus on concrete, real-world explanations for observation and events. Science starts with the important assumption that all natural phenomena can be explained in terms of identifiable materials and processes, and then generates solid repeatable mechanistic conclusions. We understand that scientific research can be conducted in a proper manner to logically and solidly distinguish among alternatives. There is an elegant beauty in this approach to investigation and conclusion.

The value of the scientific method can be particularly appreciated when it is contrasted with past ways of thinking when people had much less understanding of consistent patterns and predictability in nature. Events were as easily explained by the actions of gods, spirits, or demons as by natural causative events. Today, scientific progress has led to incredible advances in health and technology and has led to prosperity that was not even imaginable a hundred years ago.

However, something has been lost in the advancement of science. In many ways we have moved to a place as a culture where our scientific understandings and mechanistic descriptions have become all we rely on and believe in. We see that scientific knowledge has provided so much that we tend to trust it will provide the answers for everything. Ironically, the original advancement of naturalistic study was based on the premise that God created an ordered universe with predictable and repeatable parameters and processes (natural philosophy). In a fascinating twist of thinking, many people now reason that since we have increasingly described the world in terms of such parameters and processes, there cannot be a God behind it or who might intervene into it. But science is a methodology that discovers only certain kinds of truth.

As a biologist, I was fully enthralled and wholly satisfied with my own mechanistic view of the natural world until I started to recognize many things that I am now convinced are beyond the descriptions of science. Many times I experienced the mesmerizing beauty of stream water tumbling over scattered boulders covered with bright green algae. I am repeatedly stunned when I look through a microscope at the intricate form and markings on various parts of a wildflower. I am overwhelmed by the magnitude of the universe when I peer into a star-filled sky on a clear cool night. What are these experiences? How can science give a cause-and-effect explanation for them? For Christians, this beauty is a mark of God's hand on his creation. When I sit in the peaceful silence of a mountain lake on a windless morning, I sense there is a presence and I am not alone. I understand how scientists can say this is foolishness because I am not describing and explaining something empirically. Even so, Christians know this is the God of the universe and they long to respond.

There are comparable experiences in other aspects of life that are beyond the reach of scientific elucidations. Music is mechanically based on notes of demonstrable sound frequencies and mathematical relationships among the notes. However, if you love music like I do, you understand how harmonizing voices can bring an enveloping warmth, how a soaring guitar solo can send a chill up your spine, or even how a song melody can be so moving as to elicit tears. These sensations are not easily explained in a scientific manner. Readers could undoubtedly fill in similar occurrences from art, poetry, literature, and accomplishments at athletic events.

I am fully aware that scientists in evolutionary psychology and neurobiology attempt to explain these human emotions in evolutionary terms, as adaptive responses programmed into our neural circuitry. These professionals argue that there were certain survival advantages for individuals who experienced appropriate emotional or religious inclinations. Approaches like this work for some behaviors. For example, love for a spouse or altruistic urges to protect one's children should be favored because these tendencies can be tied to increased survival of an individual's genes. But this rationale comes up short in explaining all emotions of beauty and wonder.

In a similar vein, individuals have written about our common human moral compass and sense of good and evil that defy easy scientific explanations (such as our shared sense that it is wrong to murder children).[5] Again, it is increasingly popular among scientists to argue that all such experiences, from morality to beauty to a sense of a spiritual reality, are byproducts of adaptive traits that increased survival in human ancestors. However, this will always remain a speculative (and biased) framework rather than a scientific conclusion based on empirical support. Evolutionary arguments for these deep human experiences invariably involve strained series of proposed evolutionary connections that ultimately disclose that a mechanistic, scientific reality is all that is allowed. But even as neural psychologists discover certain parts of the brain as centers of religious experiences, we are left with the indistinguishable alternatives that these centers evolved either mechanistically without God's involvement or evolved precisely because this is what God intended.

This short discussion reveals that a wholly scientific worldview is actually based on an already embraced, underlying assumption that *all* human experiences must be explained exclusively using natural materialistic mechanisms. Importantly, this assumption is simply a predetermined philosophical position, a belief in particular parameters, rather than a factual statement. We will never be able to distinguish by processes of science whether a sense of "there is a God" is the result of adaptive selection or an inherent desire for a God that really exists.

The point, of course, is that there are experiences in life pointing to the existence of something beyond the mechanical aspects of the world described in scientific investigations. Science proceeds on the assumption that only natural causes are at work. This should not be turned around to say

5. Lewis, *Mere Christianity*, 17–39; Keller, *Reason for God*, 127–58.

science proves there are only natural processes in the world! The latter is a philosophical presupposition rather than a conclusion from science.

I offer the above rationale as a personal window into a spiritual or metaphysical realm beyond scientific description. I work within a philosophical worldview that accepts that spiritual elements exist beyond the natural world. There are certainly other paths to Christian faith, from philosophical arguments for the existence of God to personal convictions raised by words of the Bible. My simple intent is to demonstrate how legitimate scientists can also be spiritual or religious. This discussion also illustrates that spiritual or religious faith is not what some atheist writers like to describe as "believing something without good reasons to do so."[6]

In fact, I must argue for a crucially important understanding that *all* individuals make significant decisions about spiritual or metaphysical questions that are *based on philosophical assumptions* rather than pure empirical information. We very often refer to our choices regarding these fundamental assumptions as decisions of *faith*. This does not mean blind acceptance of a particular position, rather, it is reasoned belief in something that cannot be empirically proven. Those who follow specific religious faiths incorporate spiritual reasoning into their full view of what is going on in our existence. However, even atheists, who are "nonreligious," claim a metaphysical belief system that defines them. Upholding the assumptions that there is no God and that our existence is based only on mechanistic processes requires a faith in that position. At its core, this is no different than religious faith.

Atheists invariably object to this depiction and would rather define their position as distinctly different. But this is an unfair elevation of a viewpoint that is no more rational or philosophically defensible than religious positions. The vast majority of Americans reject atheism precisely because they do not see reasons to accept the required philosophical assumptions. The argument that atheism is unique creates a false dichotomy that atheism is based on rational data whereas all religions are based on irrational beliefs. Not only is this untrue, it is insulting to those of us who possess a reasoned vibrant religious faith that informs our lives. Instead, there is much to be gained from the common acknowledgment that we all live with rationally chosen worldviews that have significant implications and consequences.

The Impact of Scientists

It is from this vantage point that scientists can mend some of the damage done from past creation-evolution debates. Scientists should never present the

---

6. Pinker, "Less Faith."

discoveries of their discipline as if they are all-encompassing truths about our existence. It is not unusual for evolutionists with atheistic leanings to communicate that the notion of a God-Creator is a construct for weak individuals who need religion. How can this be acceptable when the vast majority of listeners believe there is a God? Instead, it is important that scientists properly communicate that all of us make faith decisions that are beyond the scope of science. It is really only from this place of common ground that we can openly acknowledge and respect one another. I urge all fellow scientists to value the reasoned faith-based components of people's lives while we simultaneously pursue the mechanistic methods of our discipline.

Even if atheistic scientists do not respond well to this appeal, I also have purely pragmatic reasons for scientists to increase their awareness of spiritual perspectives. That is, if we really want to effectively convey the details and fascination of scientific discoveries, we must become more aware of and sensitive to the views of the audience to which we are communicating. This is especially true for evolutionary scientists, given that the majority of individuals in American society hold God-centered viewpoints. An August 2005 *Newsweek* poll[7] found that 80% of individuals believed the universe has been "created by God." The Gallup polls mentioned earlier show that over 85% of Americans believe God was involved in some fashion in the derivation of life forms.

I believe scientists and the media largely underestimate the pervasiveness of conservative creationist viewpoints among both well-educated professional men and women and untrained laypeople. I have mixed with individuals who embrace these understandings everywhere I have traveled and lived in the Pacific States, from small rural towns to high-tech regions and suburbs of major cities. Given this backdrop, there is little value in nonreligious persons with evolutionary views isolating themselves from spiritual discussions. The most common approach among scientists is to assume that the weight of scientific evidence for evolution, which they accept so wholeheartedly, ought to be enough to bring Christians to understand their position. Clearly, this has not occurred over the last 150 years (!) and there is little reason to expect that any new evolutionary evidence, by itself, will bring a different impact. Something is sorely missing. I contend that evolutionary scientists will have the most meaningful and useful communications when they demonstrate an understanding of different spiritual views and the inherent need for most individuals to integrate evolutionary discoveries into religious perspectives. One thing that will transform creation-evolution conflicts is for scientists to more generally acknowledge

7. "Spirituality in America," 48-49.

that Christians are not irrational individuals who simply need to study evolution more completely.

This understanding is particularly important for science educators. As a scientist and educator myself, I want my students to not only grasp information but to also appreciate critical thinking and the value of scientific thought. To do this effectively, I must relate what I am unfolding for students to their current experiences and understanding. I lessen my chance to explain the fascinating world of biology and evolutionary theory if I am unable to show familiarity with and sympathy for the various worldviews of those who are listening. Unfortunately, the typical approach in science education today does not embrace this dynamic. In fact, teachers frequently insist on ignoring the spiritual faith of students within the artificially isolated confines of the "science classroom." This inadvertently is a major contributor to the creation-evolution divide perceived by so many. A more nuanced and thoughtful posture would be greatly beneficial and I suggest that the scientific community can indeed find ways to communicate more effectively. I will discuss this situation further in later chapters.

I truly believe that the clash between "creationists" and "evolutionists" can dissipate into a less-controversial issue as people better understand their underlying expectations and consider the consequences of current behaviors. An open dialogue takes a willingness to listen and an allowance for differences of opinion. It also requires rejection of pre-established notions and an honest readiness to alter one's perspective if given good grounds to do so. I have provided reasons in this chapter why the growing conversation should include a wide spectrum of individuals; if any of these reasons have resonated with you, I invite you to read on.

3

What Are We Talking about?

WHAT DOES "CREATIONIST" MEAN? And what is someone thinking when they use the loaded term, "evolution?" Those familiar with the creation-evolution topic know that there can be a lot of confusion when individuals use a term that doesn't mean the same thing to all listeners. Please allow me to explain what I'm specifically addressing and provide a framework for upcoming chapters. The discussion will not only explain specific topics and terms, but also reveal crucial and undoubtedly surprising distinctions.

The Terminology of Creation

The term "creationist" desperately needs clarification. A creationist is some-one who believes God is the sovereign Creator of the universe. This means *I* am a creationist! Although the term is widely applied to mean only indi-viduals who believe creation occurred relatively recently over six 24-hour days, we must emphatically reject this narrow usage as unfortunate and misleading. It ignores the many individuals with deep Christian faith who believe that the earth is older than several thousand years, as well as those who see God's hand and evolution as compatible conceptualizations. Most important, the narrow use of the term, "creationist," incorrectly implies that there is only one intellectual position that includes God.

This is an incredibly important beginning point. The proper use of "creationist" untangles someone's straightforward belief in a Creator-God from the individual's specific belief of how God enacted his creative powers.

Please note this distinction from the outset. "Creationist" means anyone who believes God brought about the natural world, regardless of how one believes God accomplished it. It follows that we must then distinguish among different types of creationists.

In *young-earth creation*, it is believed that God's creative actions occurred over six 24-hour earth days some thousands of years ago (usually from 10,000 to 15,000 years ago). A key aspect of this view is that God created each life form or "kind" by direct "special creation." Although young-earth creationists acknowledge that some natural modification in organisms occurs over time, this is always distinguished as being limited to within created kinds (e.g., all cat-like animal species belong to a single kind). This type of change is referred to as "microevolution." If a clarification is not made, reference to creationism likely implies young-earth creation. A simple internet search of "creationist" will also demonstrate that this form of creation is the most prominent in public discussions. I will refer often in this book to the anti-evolution arguments expressed by individuals with a young-earth creationist viewpoint.

Young-earth creationists are known for their promotion of "creation science" or "scientific creationism." Proponents argue that natural physical evidence can be used to test and demonstrate the proposition that God created all living things over six days some limited number of years ago. As such, the young-earth conceptualization has been offered as an alternative scientific explanation alongside evolutionary theory. I will discuss the issues involved with creation science in chapter 8.

Another creationist formulation is *old-earth creation*. In this view, creation occurred over very long epochs or ages corresponding to the billions of years implied by cosmological information and the geological record. The precise nature of God's involvement in the appearance of living organisms varies among proponents. In all cases there is a belief in significant special creation events in which God created whole life forms and/or complex biological structures. Some envision numerous progressive series of specially created life forms over geological time while allowing for relatively little evolutionary change ("progressive creation"). Others accept considerable amounts of evolutionary transformation. In either case, acknowledged evolutionary changes are distinguished from the full "macroevolution" understandings of evolutionists (i.e., common descent and ancestral transitions among all living things). Old-earth creationists are actually fairly common and various forms of the position have been around in the Christian church with significant followings since the 1800s, even though this is not well-known by the general public.

"Intelligent design" proposals, that suggest God or some other unspecified designer is revealed in the details of the biological world, have largely arisen from individuals who are explicitly or implicitly old-earth creationists. Intelligent design differs from creation science in that it does not focus on supporting any specific set of events or time frames in earth history. Instead, it is a more generalized philosophical proposition: there are so many unexplainable complexities in the natural world that it should be a reasonable scientific conclusion that an intelligent designer brought them about by special creation. Arguments for intelligent design are often framed in such a way that the "designer" is not specified as the God of the Bible. Nonetheless, it is clear that the attraction for Christians is the inclusion of possible miraculous actions by God in scientific discussions of the earth's past. I will discuss the intelligent design movement in chapter 9.

I introduce here a new designation, *spontaneous creation*, as a collective description to encompass young-earth and old-earth creationist positions. Proponents of both views reject some or all the major tenets of evolution and hold to the notion that God made at least some living things or major biological structures by "special creation." The defining feature of these views is the belief that God instantly created life forms and/or complex structures from within his unique character and attributes (i.e., "spontaneously" = arising from internal forces or causes—*Webster's Dictionary*). Although spontaneous creationists vary in their understandings of the earth's age and how much evolutionary change may have occurred, they are united by a belief that the history of life on earth involved significant independent origins of biological form by God's direct and instantaneous involvement outside of normal biological processes.

Finally, *evolutionary creation* encompasses any view that suggests God utilized the modifying and molding processes of evolution over very long periods of time to create the vast diversity of life on earth. It implies that all organisms are linked through major evolutionary connections in the past. One very important aspect of this view is the understanding that God's interactions with or within evolutionary processes will not be directly detectable by scientific investigations. In other words, God is always active (as the Bible proclaims) but he has not worked through notable special creation events. I will explain in chapter 10 the biblical and historical bases for this position and elaborate on its important implications. I am an evolutionary creationist along with many scientists and fellow believers in the Christian church. For many individuals, the phrase, "evolutionary creation," may seem like an oxymoron containing incompatible, contradictory terms. The primary goal of this book is leading readers to appreciate how coupling these two terms is a wholly accurate, encompassing and positive view.

The two key titles, "spontaneous creation" and "evolutionary creation," distinguish an essential difference among creationists. Spontaneous creationists believe God created organisms and/or complex biological order by large and significant acts of special creation while evolutionary creationists believe God created all aspects of organisms by evolutionary transformations of earlier living forms.

Evolutionary Terms

"Evolution" implies so many different things to people that it is almost useless without clarification. The concept not only causes considerable confusion, it frequently carries emotional connotations. Some people think of subtle changes in genes within a population, others visualize a fish crawling out of the sea and converting itself into a land-dwelling animal, while others fear horrendous social applications of "survival of the fittest." Please allow me to carefully set some parameters.

The *general concept of biological evolution* is the notion that the diversity of life on earth came about through the alteration of organisms from generation to generation over very long periods of time. The idea specifically implies major ancestral connections among life forms. This is what Charles Darwin called "descent with modification." Those who believe in the general concept of evolution are "evolutionists" and this formulation can be referred to as *evolutionism*.

However, it is important to make a distinction between this general concept of evolution and ongoing discussions regarding the specific mechanisms by which evolution may have occurred. It is one thing to talk about *whether* major evolutionary change has happened over time, based on evidence from the natural world. It is something different to specifically explain *how* that change may have occurred. Biologists overwhelmingly accept the general concept of evolution while they spend most of their time researching and discussing the *mechanisms of evolutionary change*.

I raise this distinction between the general concept of evolution and evolutionary mechanisms because enormous confusion and misinformation arise when the two aspects of evolutionary thinking are blended together. In particular, it is frequently argued that evolution is falsified if a full description of mechanisms is not delineated. This is not a valid line of reasoning and skeptics should not use it to reject evolution. I will later present a sampling of natural evidences for the general concept of evolution (chapters 11–14) and attempt to illustrate why practicing scientists are convinced about its validity. Whether or not readers believe I will succeed, it is important to consider that it is at least theoretically possible to present numerous observations that paint a solid picture of an interconnected evolutionary past without detailing all the precise steps by which those evolutionary transformations occurred.

Evolutionary biologists today talk a lot about unraveling the mechanisms of biological evolution and in this realm there is much less certainty. They have uncovered remarkable features about DNA genes, mutations and genetic controls. They pursue detailed comparisons among living and fossil creatures. They study population dynamics, the formation of new species and different modes of evolutionary change. But no one claims to have a full understanding of the precise mechanisms of evolution over time.

Certainly an important determinant of survival in nature is "natural selection." The idea follows logically from the observation that organisms produce numerous variable offspring that cannot all survive. If survival is determined primarily by the quality of each individual's attributes in relation to the environment, then populations will consist of individuals that are best adapted to the demands presented in their surroundings. It is as if "nature" is consciously choosing which individuals will survive (giving us the phrase, "natural selection"). This process over time should continually result in groups of individuals with traits that are matched to their environmental conditions. Charles Darwin's primary contribution in *The Origin of Species* (1860) was his elaborate and detailed demonstration of the power of this process in causing biological transformation. However, Darwin himself did not think it was the only process in evolutionary change and wrote, "I am convinced that Natural Selection has been the main, but not the exclusive, means of modification."[1] I will have more to say on the topic of evolutionary mechanisms in chapter 15.

Among individuals who accept the general concept of biological evolution, the big question is how it is incorporated into personal worldviews. Some assume that materialistic processes inherent in the natural world are all there is to existence. They believe biological evolution has resulted from these processes alone and that evolutionary mechanisms have operated without any influence from God. This view is *mechanistic* or *naturalistic evolution.* Such a perspective also implies an atheistic or agnostic worldview that is variously labeled *materialism* or *philosophical naturalism.*

In contrast, many evolutionists believe God is necessarily the creative entity behind evolutionary change. This position is commonly called *theistic evolution.* You may notice that this is the same thing as evolutionary creation. Both labels designate someone who is simultaneously a creationist and evolutionist, accepting both God as the creative force and evolutionary changes as the process behind the diversity of life. Some theistic

---

1. Darwin, *Origin*, 7.

evolutionists suggest God simply created the initial physical parameters and processes in nature and that these then took their course to necessarily lead to biological evolution. More appropriate from a conservative Christian view, others (including me) visualize God actively involved throughout the natural course of evolution.

Although evolutionary creation and theistic evolution are titles that might be used interchangeably, I strongly prefer the former and purposely avoid the latter. "Theistic evolution" implies that the emphasis or focus of the position is evolution and that God is simply a secondary modifier of the key evolutionary process. More important, the emphasis on evolution in the title sets up an undue contrast between Christians who accept evolution and all other Christian views that are referred to as "creationist" positions. This plays to an artificial distinction that unnecessarily perpetuates the creation-evolution controversy. As a Christian, I am first a creationist! I believe God is the sovereign and powerful creator and sustainer of all things. In this respect, I am united with spontaneous creationists. What separates our positions is the *process by which God created*, not a belief in evolution versus a belief in creation. This means that the different creationist approaches are best distinguished by the modifying adjective that specifies the mode of creation. The key distinction is that evolutionary creationists believe God fully utilized evolutionary processes while spontaneous creationists believe God utilized significant special creation events.

Although it is possible to designate more subtle differences among creation and evolution viewpoints, the categories I have described here capture the primary distinctions. The various positions are summarized below in Table 3.1. I will use this framework through the remainder of the book.

The Origin of Life

One topic frequently included in discussions of evolution is the origin of first life on the earth's surface. Although this issue also raises significant spiritual and biological questions, it is not directly related to modern formulations of biological evolution. Evolution is the idea that organisms can be altered over successive generations. Its prerequisite condition is living, reproducing life forms that generate multiple variable offspring. One of its defining features is ancestral continuity. Most discussions of evolutionary mechanisms rely on the survival of organisms with the most favorable traits and then transmission of these traits to the next generation. Although "chance" is one recognizable aspect of evolution (see chapter 5), biologists primarily view evolutionary change as a non-random molding process (e.g.,

Table 3.1 — The main philosophical positions regarding the diversity of life on earth.

1. **Spontaneous Creation:** God has acted through significant "special creation" events. He created living things and/or complex biological structures instantaneously (i.e., "spontaneously") from within his unique character and attributes.

    a. **Young-earth Creation:** God's creative actions occurred over six 24-hour earth days some thousands of years ago, typically believed to be about 10,000 to 15,000 years ago. Evolutionary modification in organisms is limited to within created "kinds."

    b. **Old-earth Creation:** God's creative actions occurred over the billions of years implied by cosmological information and the geological record. Varying amounts of evolutionary transformation are accepted among different old-earth positions; in all cases, it is believed that they are inadequate to account for the diversity of life on earth.

2. **Evolutionary Creation (Theistic Evolution):** God created living things by utilizing the modifying and molding processes of biological evolution over very long periods of time.

3. **Mechanistic Evolution / Philosophical Naturalism / Materialism:** Materialistic processes inherent in the natural world are all that exist and these have brought about the diversity of life on earth. Evolutionary mechanisms have operated without any influence from God, gods, or other spiritual entities.

natural selection). The key point is that these processes are acting within a system of already-existing and perpetuating life.

It is a separate thing to ask where the very first living organisms came from. The origin of the first cell (or cells) capable of maintaining form through metabolism and also capable of making replicates of itself requires explanations of how such complexity can arise from inorganic precursors. Although there could be selective mechanisms in this event that parallel natural selection, the discussion revolves much more completely around how non-directed chance events could produce the complexity of the first living forms. In other words, the topics of evolutionary change and the origin of life are different in a fundamental way. Biologists are a long way from satisfactory explanations for how the latter event might have occurred. Once again, Darwin himself recognized the difference between the origin of life and evolution itself. He wrote as such in the closing lines of *The Origin of Species*:

> There is grandeur in this view of life, with its several powers, having been originally breathed by the Creator into a few forms or into one; and that, whilst this planet has gone cycling on according to the fixed law of gravity, from so simple a beginning endless forms most beautiful and most wonderful have been, and are being, evolved.[2]

The distinction between evolutionary change and the origin of first life is important. Some authors have taken the approach that if they are able to reject scientific explanations for the origin of life, they have also brought down the foundations of evolutionary theory.[3] This is not a valid rationale and should be rejected as an overly forced association of topics that are based on some different assumptions and different arguments. Although I acknowledge that there are very important questions in the issue of life's origin, my focus is on something different. The topic I discuss throughout this book is biological evolution.

2. Darwin, *Origin*, 490.
3. Strobel, *Case for Faith*, 87–112.

PART II
Real Issues for Christians

READERS MIGHT GUESS THAT I would get right to the specifics of biological evolution and explain the sorts of things that convince me and other biologists of its validity. Many years ago, I was among those who believed the creation-evolution controversy could be resolved if biologists were simply given a chance to carefully explain the evidence for an evolutionary view. I have come to understand that the issue revolves much more around cultural, philosophical, and theological implications of evolution rather than any specifics within biology. Both Christians and non-Christians are usually most eager to discuss these implications. In fact, people often bring a strong need for them to be clarified before a meaningful look at biological information can happen. In addition, people often carry relatively unexamined assumptions and opinions that greatly influence how they view the creation-evolution topic. Because of these real phenomena, the next several chapters address areas of concern that are regularly raised about biblical creation and evolutionary thought.

I begin here in Part II with prominent topics for Christians, especially for evangelical believers. I acknowledge upfront that I aim to question a number of assumptions that are widely accepted within the church. I expect that many Christian readers will find portions of these chapters challenging or even unsettling and I recognize that this is inherently uncomfortable. Nonetheless, it is a necessary exercise to effectively reveal how creation and evolution are wholly compatible concepts. For readers who find these issues are not personally challenging, I encourage you to use these discussions to gain a deeper understanding of the mind-set of those for whom they are.

4

What Can the Natural World Tell Us?

The natural world is a wondrous symphony of intricate elements. Scientific investigations attempt to uncover the mysteries of all the participants and their complex interactions. In general, we appreciate and even marvel at what is revealed. For example, scientists have precisely described how increasing daylight and rising temperatures during the spring trigger internal mechanisms within plants so that they bud new leaves filled with bright green chlorophylls and other pigments. We also understand how the same cues stimulate behavioral urges in many bird species so that they navigate hundreds of miles by use of the night sky to return to their traditional breeding grounds. Christian naturalists were historically at the forefront of scientific discoveries like these as believers considered it an honorable endeavor to reveal God's glory in the intricacies of his natural creation. Today, investigations in medicine and other areas of contemporary application are especially valued for their positive contributions to human health and our quality of life. However, our generally shared, positive human response to scientific studies is disrupted when the subject matter is ancient earth history. Many Christians, of course, bring biblical understandings that conflict with descriptions in the historical sciences. This is a phenomenon that is not generally encountered in other disciplines. The critical topic of this chapter is how Christians respond to perceived contradictions between the words of Scripture and scientific investigations of nature.

It Never Really Happened

A good number of Christians believe they have a straightforward solution for the confusion over creation and evolution. They propose that God created the natural world with attributes that make it appear to be very old even though the earth is actually quite young. In this approach, one argues that reported physical evidence for evolution and a long earth history is not reliable simply because the Creator formed everything with a false impression of past existence. An often-mentioned illustration is when God created a forest of large old-growth trees by special creation, each of the trees contained annual growth rings in its wood suggesting an age that was not, in fact, a reality. Similarly, the light streaming toward earth from a star one million light years distant was created along with the star itself. It follows that God could have created a variety of natural features that appear to suggest evolution had occurred but these features wouldn't truly represent an evolutionary past. This line of thinking may seem reasonable at first, but the suggestion carries with it at least two negative and problematic implications.

First, it means that nothing can truly be known about the past. For some conservative Christians, this leads to a clean and simple conclusion that the Bible is the only reliable source for knowledge about historic events. But it also suggests that all or a large part of many scientific disciplines, including cosmology, astronomy, physics, geology, and pre-biblical archaeology, are based merely on illusion. This might be satisfying for some Christians, but it is important to consider that from all other vantage points it amounts to a strange anti-intellectualism. Should we really conclude that nature does not mean what it appears to mean in all these scientific realms? This is certainly not beneficial to a church that claims it is offering truth to the world. If God gave all of us a mind for thinking and reasoning, shouldn't it be used to study the natural world?

The idea that Christians can base all knowledge about past events on the Bible alone is only a relatively modern notion in the church. It is counter to a strong Christian tradition that God speaks to humans through both Scripture and nature. John Calvin, the influential Reformation theologian, taught that believers should consider knowledge about nature, the works of God, an important aspect of knowing the Creator. He wrote, "Let the world become our school if we desire rightly to know God."[1] To most observers the "appearance of age" argument is a retreat from the scientific community rather than a helpful solution. If evidences in the natural world are simply appearance, there can be no meaningful scientific arguments about the his-

---

1. Calvin, *Commentary*, Vol. 1, Argument.

tory of life on earth. Even so, can the church really claim a victory with such a position?

A God of Deception

A second and perhaps more crucial implication of the false appearance argument is what it suggests about God, the Creator. It implies that his creation is full of deception—not just a little, but detailed and elaborate deception (remember, this position means that the scientific fields listed above are based on enormous amounts of intricate faked information). I am fully aware that Christians must be careful before we claim to know what God would or would not do, but this is entirely counter to God's character as revealed in the Bible. As Catholic evolutionary biologist Kenneth Miller notes, using the "appearance of age" approach means that "we must simultaneously conclude that science can tell us nothing about nature and that the Creator to whom many of us pray is inherently deceitful."[2]

The Bible proclaims that God does not lie[3] and statements of "I am the truth" and "I tell you the truth" characterized Jesus' ministry on earth. Scripture teaches believers to put their trust in God and to desire the truth.[4] Deception is the antithesis of God and is equated with evil.[5] The Bible also teaches that God's creation reveals truth.[6] According to Psalm 19:1–2, "The heavens declare the glory of God, the skies proclaim the work of his hands. Day after day they pour forth speech; night after night they display *knowledge*" (my emphasis). Christians are called to trust in a God who is wholly truthful and faithful, not one who would create a deceptively fake image of reality in his natural creation. The important implication is that God reveals truth both in the Bible in *and* in nature. It may be a foreign concept for many Christians but if science appears to conflict with a biblical understanding of earth history, we must be drawn to reexamine our interpretations of both the scientific findings *and* the biblical text. A balanced scriptural view trusts that honest scientific study and open analysis of the Bible will be wholly compatible with one another.

Some young-earth creationists go so far as to claim that God is simply testing our faith by making us choose between the Bible and "the world." They advise Christians that the true nature of our faith is at stake and that

2. Miller, *Darwin's God*, 80.

3. Titus 1:2, Heb 6:18.

4. John 14:1, Ps 51:6, many other Psalms.

5. Gen 3:13, Lev 19:11, Matt 24:5, 2 Thess 2:9-10, 1 Pet 2:22.

6. Ps 50:6, 85:11, 97:6, Job 12:7-8, Rom 1:20.

we must make the choice to stand up and appear "foolish" before the scientific community. This admonition is based on 1 Corinthians 3:18–19: "If any one of you thinks he is wise by the standards of this age, he should become a fool so that he may become wise. For the wisdom of the world is foolishness in God's sight." This may seem to be a position that honors God, but the message to the Corinthians was that the world might view the *spiritual truths of the gospel message* as foolishness, not that all wisdom of the world is unreliable or that there is value in appearing foolish pertaining to it. Another verse in 1 Corinthians (1:18) clarifies: "For the *message of the cross* is foolishness to those who are perishing, but to us who are being saved it is the power of God" (my emphasis; also see 1 Cor 1:20–25). I encourage Christians to make a careful distinction between standing up for faith in Jesus and alienating nonbelievers by holding on to a dismissive position regarding origins. If believers use the "appearance of age" argument, the church is sadly carried further away from the time when Christians were leaders in intellectual communities, including within the scientific disciplines.

The issue I raise here about appearance, deception, and God's character will repeatedly emerge when I discuss the evidence for evolution. I will try to demonstrate why an accumulation of physical signs in the natural world has persuaded the biological community to accept the general concept of biological evolution. An honest biologist will also admit that there are many things that we don't understand. There is frequent quibbling in the debate over creation and evolution about specific examples of evidence, but the real point for evolutionists is precisely that there are so numerous and varied *appearances* of evolutionary change about which to quibble!

I can tell you ahead of time that no discussion of natural evidence will paint a complete picture of exactly what has happened in the past. Instead, it is like having a good number of pieces of an incomplete puzzle. Ultimately, the acceptance of evolution or alternate ideas will be based on what we can make out in this puzzle. I will argue for an evolutionary creation view because there is so much information that supports the overall notion of biological evolution over very long periods of time. If spontaneous creationists acknowledge any part of this pattern of evolutionary evidence, they are necessarily faced with truthful meanings in the natural world or they must deal with the issue of false appearance that would seem to be inconsistent with God's character. If we accept that God would not deceive us in such a way, the underlying question for a Christian skeptic is: how much evidence for (or appearance of) evolutionary change is enough (even without all the details) before he or she is compelled to recognize evolution as a real phenomenon?

One example where this problem of implied deception arises is with an old-earth creationist approach referred to as "progressive creation." Proponents accept that ancient rocks and fossils tell a factual story about different periods over millions or billions of years of earth history. However, they do not believe that the myriad animal and plant forms came about through evolutionary transformations. Instead, they suggest that God produced new forms by spontaneous creation in a long progression of creative events; as some species died out, new forms were created to take their place. This old-earth position is attractive to some in that it explains what is observed in the earth's fossil record while preserving the notion that God created life forms by instantaneous special creation. The difficulty for progressive creationists is they also acknowledge that many fossil forms appear in series or sequences consistent with evolutionary progressions.

Progressive creation relies on the argument that all specially created species necessarily go extinct due to processes of decay and decline (with reference to the second law of thermodynamics) and their inability to adapt to changes the way evolutionary biologists claim.[7] Therefore, sequences of species in fossil accounts must be the result of successive creation events by God. One problem is that this proposal misapplies thermodynamic principles in living things, which by very definition counter the effects of entropy by capturing nutrients and energy to maintain form, grow, and produce progeny. Also, there is continuity in life from parents to offspring and no inherent reason that entire species units should be only temporarily viable. Even if we put these arguments aside, progressive creation forces us to ask why specially created species should be produced in sequences that surely would be interpreted by modern scientists as the product of evolution. Since God could have created new species in *any array of forms* entirely unrelated to previous forms, we are faced with a clear problem of creation containing false appearances of an evolutionary past. This implies, in turn, elaborate deception by God.

The Bible Comes First

If we proceed with the assumption that God does not deceive us in his natural creation, there is a second related issue of enormous importance. A very common proclamation among evangelical Christians is that Scripture must be given precedence over evidences in the natural world. That is, if there are disagreements between the Bible and biology, Christians should clearly favor God's inspired word and judge the information from nature as faulty. The

7. Ross, *Genesis Question*, 47–58, 155.

way this argument goes, it isn't that God put false appearances in the physical world but it is our fallibility as humans to accurately examine the evidence that leads us to false conclusions. As one Institute for Creation Research associate put it, "the Bible-believing Christian recognizes that his unaided mind and faculties of observation cannot solve the basic problems dealing with earth history. The Christian trusts in a revealed record from God, Himself" or as other young earth-creationists have stated, we "must grope in darkness apart from God's special revelation in Scripture."[8] This is a widely held perspective among Christians in America today and it dominates the culture of evangelical churches. However, I am convinced that most believers have not been shown the full nature and implications of this approach.

I understand the high authority of the Bible, as it is God's revelation upon which Christians depend for so many truths regarding who God is, our relationship to him, and how we ought to live. I am among those who believe regular reading in the Bible is an essential activity for Christians so we may renew our minds, grow, and be transformed (Rom 12:2). The primary problem with the above argument is that it presumes there is only one correct view about what Scripture says pertaining to earth history, usually six-day, young-earth creation. This is the view we are supposed to favor over faulty physical evidences. But many disagreements occur over human interpretations of the Bible as indicated by the numerous Bible translations and commentaries, each of which attempts to present the best possible understandings. How are we supposed to favor what Scripture says if we don't agree on what that is? No matter how strong a theological case someone believes he or she can make for a young-earth viewpoint, it is only one possible reading of the Bible. A broader consideration reveals that the "days" of creation have never been universally interpreted among theologians as 24-hour periods (explained in later chapters) and the Bible does not clearly communicate how God performed his creation actions. Instead, Christians who truly cherish the Bible as the inerrant Word of God hold a diversity of creationist positions, including old-earth creation and evolutionary creation.

There is a crucial distinction here between the inerrancy of Scripture itself and the supposed inerrancy of any particular interpretation of that Scripture. If there are good natural evidences to support evolution, or a very old age for the earth, then to claim we should defer to a young-earth interpretation of the Bible is really an emotional appeal inappropriately fortified by reference to our high view of Scripture. A young-earth creationist interpretation is not equivalent with the Bible itself. I encourage Christians,

---

8. Nevins, "Earth History"; Whitcomb and DeYoung, *Moon*, 69.

especially those with evangelical backgrounds, to ponder the consequences of this statement.

A strong appeal to biblical authority to reject evolution is a significant departure from the guidance of prominent Christian thinkers of the past. Many voices have warned that, although we should deeply treasure the Bible, we should be very careful about what we say it proclaims with regard to scientific information. This message stretches from Augustine in the 4th and 5th century through Thomas Aquinas, John Calvin, and others. Aquinas wrote the following in the midst of discussions during his day about the nature of the earth, sun, planets, and stars:

> Two rules are to be observed, as Augustine teaches. The first is, to hold the truth of Scripture without wavering. The second is that, since Holy Scripture can be explained in a multiplicity of senses, one should adhere to a particular explanation only in such measure as to be ready to abandon it, if it be proved with certainty to be false; lest Holy Scripture be exposed to the ridicule of unbelievers, and obstacles be placed to their believing.[9]

As I have noted, an overwhelming majority of scientists are convinced of the validity of evolutionary theory based on natural evidence while practicing scientists who are spontaneous creationists represent a very small percentage at the fringe of the scientific community. It is, therefore, no small thing for a Christian to claim the conclusions of scientists are faulty because they conflict with a favored reading of the Bible.

An especially important observation is that the "primacy of Scripture" approach is a relatively modern conceptualization that is inconsistent with historical orthodox teachings of Christian faith (see chapter 6). God's created world (often referred to as his "general revelation") is a reflection of the Creator and Christians have always been intellectual leaders in the pursuit of knowledge pertaining to it. It has been commonly understood within the church that the natural world can inform our understanding of God's written words in Scripture. For example, the well-known 19th-century theologian Charles Hodge (1863) wrote:

> The proposition that the Bible must be interpreted by science is all but self-evident. Nature is as truly a revelation of God as the Bible and we only interpret the word of God by the word of God when we interpret the Bible by science.[10]

---

9. Aquinas, *Summa*, First Part, Question 68.

10. Noll and Livingston, "Charles Hodge," 66.

Something very different is happening when evangelical Christians rely on a predetermined young-earth biblical interpretation and in so doing inadvertently or purposely withdraw from many academic arenas of the modern biological community.

Are Presuppositions the Key?

The primacy of Scripture argument is utilized in a distinctive fashion within the messages of the two most influential young-earth organizations today, Answers in Genesis (AiG) and the Institute for Creation Research (ICR). A core assertion is that all truths about the earth's past are based on "presuppositions."[11] That is, all scientists are only capable of interpreting facts from the natural world in a manner consistent with prior underlying assumptions. For AiG and ICR, a six-day, young-earth scenario is presupposed according to "biblical authority" and the "infallible revelation" of the Creator in the Bible. These groups argue that modern scientists have an equal countering presupposition that the earth is old and that major evolutionary transformations have occurred in the past. As Ken Ham, president of AiG, proclaims, "The battle is not about the evidence or facts . . . it is how you interpret the facts."[12] For followers of AiG and ICR, evidence will always confirm a young-earth account if properly understood and interpreted through the lenses of their correct presupposition. Any other conclusion from biological information by fallible humans is suspect in lieu of the infallible words of the Bible.

There is an incredibly significant and disturbing implication in this presuppositional argument. It means that natural evidence has no real value to inform us apart from what we already believe to be true. It means our differences will in no way be altered by scientific discoveries. In other words, the creation-evolution issue is reduced to nothing more than a debate over personally chosen philosophical worldviews!

Although the presuppositional framework may sound persuasive, it is flawed with enormous logical inconsistencies that I ask Christians to carefully evaluate. We cannot logically reject an open study of natural information because we understand humans are fallible and then ignore the necessary correlate that it is the same fallible humans who are attempting to interpret the Bible. If we cannot be trusted with correct interpretations in science, why should we think we can be trusted with interpretations of the

11. Ham, "Young Earth"; Ham, "Proof of Creation"; Jeanson, "Presuppositional Research."

12. Ham, "Magic Bullet."

biblical text? Some Christians may claim that the Holy Spirit will open our eyes to the words of Scripture, but there are no solid grounds for claiming that the Holy Spirit cannot lead believers just as faithfully in interpreting God's natural creation. Further, if natural studies of evolutionary scientists are to be dismissed as fallible and unreliable, then all interpretations of nature by young-earth creationists should be treated as equally incorrect and irrelevant. The truth is that no human is in a position to claim that he or she knows for certain that the Bible describes a six-day, young-earth, special creation event and we should not allow honest investigation about the natural world to be dismissed as mere belief or personal choice.

A frequent charge of spontaneous creationists is that the scientific community is being dishonest because of its own predetermined evolutionary presuppositions and atheistic agenda. They argue that evidence for evolution is discovered in the natural world mainly because scientists begin with the belief that it will be found. I want to be clear and acknowledge that modern biologists *do* proceed with an evolutionary presupposition. However, it is based on a long history of previous evolutionary discoveries, not a predetermined worldview. I will later discuss undue biases and atheistic leanings that improperly influence the pronouncements of many biologists (see chapter 7). Nonetheless, the charge that evolutionary science is only a chosen presupposition driven by atheism is an imaginary scenario with no real foundation.

Modern evolutionary theory has developed from a history of research and strong challenge for over 150 years. Most 19th- and early 20th-century naturalists had a heritage of special creation beliefs, but they were persuaded to an evolutionary view by a preponderance of natural confirmations in God's creation. Today, there are far more Christian biologists who accept evolutionary theory than Christian biologists who hold to spontaneous creationist beliefs. This cannot be explained away (as is often attempted) as believers simply falling away from core principles of Christian faith. Atheism cannot be the motivating force behind evolutionary theory when so many Christian scientists favor it. Scientists expect that new findings will add to previously established understandings of evolution because the latter have developed from a large and integrated collection of prior discoveries. That is, we are proceeding with a *scientific* presupposition based on a long history of experiences with the natural world. This is very different than imposing a *biblical* presupposition onto scientific endeavors and claiming we can make all possible scientific information mean what we want it to mean.

The most concerning attribute of the young-earth claim to biblical authority is an explicit admission that there is no reasoning from fellow Christians or the scientific community that might allow a believer to move outside a six-day, young-earth framework. A closer look exposes the intellectual confines of such a position. Ken Ham writes, "Just because we don't readily know what may be wrong with man's theories doesn't mean God's Word is untrustworthy. It just means we don't have all the information to know why man's ideas are incorrect."[13] How about the obvious alternative that God's word and scientific discoveries are both reliable and it is a young-earth Bible translation that is untrustworthy?

Further, leading young-earth creationist writers reveal that their "research" is limited to only two kinds of endeavors.[14] The first is to "perform experiments to fill in details that the Bible omits" and the second is to "critique published data and reinterpret it as necessary" (that is, to make it comply with a young-earth worldview). Why can we not instead openly evaluate scientific information based on it's own merit to see where it leads?

Finally, a common emotional claim among young-earth creationists is that using natural evidence to dispute their position is elevating the importance of extra-biblical sources over the infallible words of Scripture. That is a false portrayal. As believers in a sovereign God of creation, Christians should fully expect that nature and the Bible will complement and inform one another; this does not elevate the former over the latter. However, natural evidence can, and should, be elevated above any person's *interpretation* of the Bible if there are major conflicts between the two. This does not lessen Scripture; it honors and confirms it!

Current Conditions for Evangelical Christians

From my vantage point within the conservative evangelical church, it is remarkable and unsettling how deeply the primacy of Scripture approach has penetrated. Prominent communications support the notion that evangelicals already know what we believe about earth history based on detailed biblical analysis. Many evangelicals accept that a six-day, young-earth biblical interpretation is the only valid reading of God's Word. A common corollary message is that there is no way scientists can prove biological evolution or, even more pointedly, that they hide the fact that what they claim is based on "pure speculation" or "vain imaginations." Sadly, I have found that the majority of Christians who embrace these perspectives have little understanding of what modern biologists are really saying or how natural observations might possibly be reconciled with a young-earth position.

13. Ham, "I Know Nothing."
14. Jeanson, "Presuppositional Research"; Morris, "Creationist Research."

One thing to which I can certainly attest is that within many conservative evangelical circles there isn't much room for questioning or discussion of alternatives. In the most severe communications, a Christian's saving faith in Jesus is directly coupled with acceptance of a young-earth biblical view.[15] Some may think or suggest that the latter is only an extreme circumstance, but I believe large numbers of evangelicals can confirm that an association of young-earth creation with the central message of salvation in Christ is often implied if not explicitly stated.

Given this backdrop, one can readily understand the fundamental clash in publicized aspects of the creation-evolution controversy. The underlying young-earth creationist message is that the authority of the Bible automatically trumps all biological discoveries that conflict with it. This means any new information reported by the scientific community that adds support for evolution (such as a new fossil of a feathered dinosaur) provokes an immediate rebuttal from young-earth authors who are admittedly committed to the notion that it cannot be accepted. It is from this mind-set that many Christians dismiss not only evolutionary information, but also those who acknowledge it. It is indeed a strange sensation to be personally rejected on this basis by my brothers and sisters in Christ.

But how can there be such confidence when devoted, conservative, scholarly Christians from the past and today do not even come close to a united agreement with a six-day, young-earth biblical interpretation? It is difficult to visualize our Lord saying "well done, good and faithful servant" for vehemently defending a cherished interpretation of Scripture while those whom we are supposed to reach with God's redeeming message have good reasons to understand that this defense is offensive and evasive. I suggest that a more balanced and reasoned process for believers is to openly and prayerfully study both the words of the Bible and the vast array of information from the natural world for what each can reveal to us about the God who is the author of both. Is this not a more full expression of faithfulness to the God and Creator whom we worship?

15. For a concrete example, see "Man Walked With Dinosaurs" by Pastor Mike Hoggard, Watchman Video Broadcast and Bethel Church in Festus, MO. The presentation includes the following:

"You, according to the Scriptures, are without excuse. If you want to believe that God can save you and give you new life, you need to believe that God created the universe exactly the way the Bible spells it out . . . The same God that you want to create in you a clean heart is the same God that you must believe in that created the universe in six days, six thousand years ago. Where is your trust going to be? Is it going to be in the vain imaginations of people who hate God or is it going to be in the purity of the word of God that has never been proven wrong one time."

Moving Forward

Before proceeding further with the creation-evolution topic, I pose two crucial questions for Christians. First, are you willing to openly evaluate and
trust physical evidences in the natural world and accept what they imply?
Second, if there are good evidences for evolutionary change and an old age
for the earth, are you willing to reevaluate biblical interpretations that are
inconsistent with that evidence? If you cannot visualize answering affirma-
tively, then this book may not be for you.

I'd like to describe how this looks for Christians when it is fleshed out.
When natural evidence is evaluated, we understand that it is something God
has chosen to reveal to us. We allow scientific discoveries to glorify God in
all possible respects. In contrast, we are not hovering to bring in Scripture
to clarify or argue that evidence cannot mean one thing or another. We
must let the natural evidence objectively lead where it leads. Then, calmly,
rationally, and with an open mind, we consider how both the Bible and
these revelations from the natural world fit together. The Bible will always
inform our scientific view by declaring that there is a sovereign Creator-
God behind everything that is studied, but we should not be surprised that
science reveals aspects of creation that were not described or hinted at in
biblical texts. This approach seems immediately foreign to someone seeped
in young-earth creation, but why should it really? This is not an elevation of
general revelation (nature) over special revelation (the Bible); it is an integrated coherent view of God's entire message to his people. The proposition
is not to compromise one's devotion to God or to be "blown here and there
by every wind of teaching" (Eph 4:14) but to expand one's understanding of
how awesome God is. It is also not to compromise the integrity of the Bible
but to fully enhance one's interpretation of Scripture to include all the glory
that God has chosen to reveal.

One of the best current resources for understanding an evolutionary
creation approach is the BioLogos Foundation (www.biologos.org), a national organization with numerous online contributions by various scientists, theologians, and pastors. I suggest that Christians with conservative
backgrounds begin with the organization's affirming "What We Believe"
document (biologos.org/about) and then explore the introductory resources offered on the home page. BioLogos wonderfully promotes humility and
respect for individuals with different opinions. This means some articles
will undoubtedly be too liberal for some evangelicals, while others will more
effectively address their concerns. In all, there is much to commend to those
who are serious about reconciling biblical Christianity and modern science.
Included among the BioLogos posts are encouraging testimonies from

individuals who express thankfulness for being freed from young-earth creation restrictions to a more expansive and satisfying understanding of God's natural creation.[16] I am hopeful that the perspectives I've presented in this chapter will help other Christians travel joyfully and fruitfully along this path.

16. BioLogos, "Stephen Pelton"; Buller, "Video Testimony"; Falk, "Joanna's Story."

5

God Wouldn't Do It That Way

IF YOU ARE A Christian, it is a significant question to ask if you carry any particular sense about the manner in which God *should* have gone about his creative activities. I do not only mean a sense about the time frame of creation, but also about the character of the process he might have utilized. For example, some young-earth creationists object to biological evolution on the grounds that they believe God would not use such a time-consuming, cruel, or wasteful process. They also argue that to accept a universe and earth that are billions of years old or the idea that God used processes of evolutionary change reduces creation to an insignificant event, implies a less powerful Creator, and suggests attributes of God that are unlikely. This chapter unwraps a few of these qualitative judgments about how God would accomplish his creation of the world. The primary complication with statements like these is they are largely based on human reasoning and/or emotional impulses limiting what God conceivably could or could not have done. Although they may at first sound very logical and appealing, biblical support for them is not very strong and they do not align at all with what the natural world reveals to us.

When we look at the known universe we see an almost unfathomable scale of distance measured in many *billions* of *light years* (a light year is the distance one would travel in one year if moving at the speed of light—incredibly far!) and an absolutely astounding number of stars (over 100 *sextillion* as a minimum estimate—that's 1 followed by 23 zeros!). These things speak to Christians of God's awesome power and our humble place compared to him (Ps 8:3–4). Why then should it surprise us that God might utilize large amounts of time that are almost beyond our comprehension to bring about his creation? Such a thing is more consistent with the mind-boggling numbers pertaining to the size of the universe and the number of stars than a creation period limited to days or even thousands of years. Is it not possible that our small human-centered time scale throws off our perspective? It is, of course, ultimately presumptuous to assume God would accomplish his creation only in a time scale to which we can relate. From what we can observe, God must certainly take pleasure, apart from us, in such an immense universe. This is true even though one that was a thousandth the size and contained a fraction as many stars would still be equally astounding to our perceptions. Bringing preconceived ideas to this issue may be hindering our ability to see what God is really like.

God's Good Creation

In the same manner, to claim that God couldn't use evolutionary processes because they are cruel or wasteful unjustifiably pigeonholes what he might or might not have done. God has established physical laws by which the natural world is governed and has allowed his creation to proceed under the operation of these laws. For example, all organisms must tap into an external source of energy to maintain their structures against the natural tendency toward entropy. Photosynthetic plants and algae convert energy from sunlight to chemical energy that they can utilize, while all other forms of life (except some unusual bacteria) must consume the body tissues or products of other organisms. There is no conceptual existence for earthly organisms without herbivore-plant and predator-prey relationships. This means that death and "waste" are a necessary aspect of creation.

The simple act of breathing implies internal activities of energy conversion that are necessary to counter the never-ending forces of molecular decay. Oxygen that animals inhale is used within cells where energy-carrying molecules (ATP) are constructed, while carbon dioxide that results as a byproduct of the process is exhaled. This familiar exchange of oxygen and carbon dioxide has no meaning outside the context of constant molecular decomposition. This, in turn, means lungs in created animals would be completely without use and irrelevant if there was not molecular decay! Furthermore, the phenomenon of reproduction (a defining characteristic of living things) has no full context without a corresponding death of individuals. Reproduction in the absence of mortality would quickly lead to disastrous overcrowding and exhaustion of limited resources.

Even the Garden of Eden must have had these principles in operation. Various theologians have suggested that there was no pain or death in the Garden and that carnivorous animals such as lions (that we would expect would kill their prey) ate grass or other plant material. However, if there is no decay or death (i.e., no entropy) then organisms have no need to either eat or reproduce. There is no scriptural support that animals ever ignored these things. Genesis 1:24–25 states that God created "livestock" (or "cattle"), which explicitly implies food for humans, and Adam is told that he could *eat* from any tree except one.

Additionally, a lion is only a lion when it is defined by its carnivorous teeth, a digestive tract designed to process meat (something quite different than in plant eaters), and a physical form adapted for quick sprints to capture prey. To say that this animal was able to eat and survive on plant material in the Garden is to describe something that is no longer a lion. Even if all animals were herbivores, death and damage to plants necessarily result. There tends to be an emotional response to the death of animals because we anthropomorphize their suffering, but the destruction and death of plants are, in principle, no different.

It is also revealing that Genesis 2:15 states that God took man and placed him in the Garden "to work it and care for it." What purpose is there for "work" and "care" if it is not in assuring water and good conditions for plants in the Garden and protecting them against opposite negative conditions that might lead to harm (i.e., death)? We can also conclude that the production of offspring was inherent in organisms from the very beginning. Genesis 1:11–12 describes God's creation of various plants with seeds, the latter being a mechanism for reproduction, and the familiar phrase throughout Genesis "according to their kind" or "after their kind" has always been interpreted as producing progeny. Finally, God tells Eve after The Fall that her pain in childbirth would *increase* (Gen 3:16), not that it would start from that point on. Pain was not absent from the Garden. It feels good to assume that God would not allow death, pain, or waste in his creation process, but the idea is not logically sound and does not hold up under a complete view of Scripture.

Biblical Objections?

Two Bible references are often used to reject the possibility of evolutionary processes that rely on pain, suffering, and death. The first is the repeated phrase in the Genesis 1 creation account, "And God saw that it was good" and then the final declaration, "it was very good." For many, the simple fact that a loving and gracious God would declare creation "good" rules out negative attributes such as pain and death. But this is a human construction of what "good" should mean, not what the text actually says. In fact, it is precisely the Genesis 1 verses around these declarations of "good" that describe a creation not unlike what we are familiar with today containing seed-bearing vegetation and livestock with direct indications that food was consumed and reproduction occurred.

It is fitting to look for more nuanced interpretations of the word "good" in the verses of Genesis 1. I will have more to say about this in the last section of the book, but one of the most meaningful messages is that God provided an orderly, intended world for the human beings he created in his own image. In this context, "good" means well-ordered, appropriate, and pleasing, but does not indicate the absence of all negative attributes. I believe it is noteworthy that Genesis 1 does not say the creation was "perfect," without pain or suffering, although that could have been stated if it was the real intent. That description in fact is reserved for the culmination of the New Jerusalem in Revelations 21:4: "There will be no more death or mourning or crying or pain, for the old order of things has passed away."

The second and probably most significant passage used to refute biological evolution is the "Fall" of Adam and Eve in the Garden of Eden. A widely communicated and well-known interpretation is that the Garden of Eden of Genesis 2 was an idyllic, perfect setting and that all physical death and suffering entered the world as a result of Adams' actions explained in Genesis 3. If this is the correct view, there could not have been any evolutionary history prior to The Fall. It is indeed intriguing that so many people believe that this is what Genesis 2 and 3 describe, as these elements are not readily apparent in the contained text.

There are various reasons to conclude that these depictions are inaccurate readings. First, there are the biological and biblical arguments just described showing that there must have been decay, pain, and death in the Garden. A straight reading of Genesis 2:8–17 indicates only a pleasing garden setting into which Adam was placed with no description of the unique, remarkable conditions often attributed to it. Second, an analysis of Genesis 3 and New Testament passages such as Romans 5:17–18 indicates that Adam's Fall led to *spiritual* death, separation from God, for all humanity, not to *physical* death. Adam did not physically die from his act of disobedience; rather he experienced the condemnation of being banished from God's presence. Through Jesus, the counter to Adam, humans are offered grace and an opportunity for a new spiritual life in a restored personal relationship with God. Even where there is implication that Adam's Fall is connected to physical death for humans (e.g., 1 Cor 15), it was through a lost unique opportunity for eternal life available only to Adam (i.e., access to the Tree of Life). There are no verses that communicate a previous condition of human immortality that was removed (see chapter 18 for further elaboration).

There is also no specific indication or reason to expect that other living things in the Garden ever existed in a state of perfection or eternal life. The best understanding from what the Bible and the natural world reveal to us is that normal biological processes, such as photosynthesis by plants and digestion by animals, as well as growth, reproduction, decay, and death, occurred continuously from long before Adam and Eve to the present day. Many young-earth creationists have insisted that the absence of animal death and suffering prior to The Fall is essential to the Christian message of redemption through Jesus, but this forces an unnecessary (and unfortunate) stipulation onto the Christian gospel that is inconsistent with a straightforward study of Scripture.

Random Processes in God's Purposeful World

All modern explanations of evolutionary change include the idea that new possibilities arise in organisms by random genetic events. Many Christians have focused on this aspect of evolution with the strong objection that God would not utilize elements of chance in his intended creation. The objection actually comes in two forms. One version is the complaint that evolutionary biologists claim the diversity of life came about "completely" by chance, as if randomness is basically all there is to evolution. This is a misrepresentation of evolutionary descriptions because it excludes or downplays the crucial, simultaneous, and even more important role of natural selection. The latter is a refining, directed process that is the antithesis of chance. Evolution cannot be evaluated meaningfully without an accurate grasp and inclusion of selective processes. I will return to this topic in chapter 15 where I discuss what biologists really know about the mechanisms of evolutionary change.

A more fundamental rejection of chance events by Christians comes from theological and emotional notions that there simply cannot be random, non-directed processes in an intended world created by God. Many skeptics argue that there is an inherent, unsolvable contradiction in trying to merge purposeless processes with a purposeful God.[1] Others suggest that to claim God allowed or utilized chance events in his creative process is an insult to his sovereignty. As with the previous topics in this chapter, a closer view of the natural world and careful consideration of what is being implied

1. Morrow, "How Should Christians."

reveals that these arguments are not based on an accurate view of what we know about God.

Scientists understand there are all kinds of random, non-directed mechanisms in the natural world, but also that these are all part of an orderly existence. One of the early topics in many biology courses is the concept of "diffusion" (some readers may remember this with either a smile or a groan). If we add a drop of dye to a beaker of still, clear water, the dye particles spread out so that, in time, the beaker contains a uniformly colored mixture of dye and water molecules. If I open a bottle of extremely strong perfume, some of the perfume molecules will eventually reach the nose of someone fifteen feet away.

So what caused the dye or perfume molecules to move from their source location? The simple answer is the continual, *random, non-directed,* jiggling motion of all molecules in solutions and gases (called the kinetic energy of molecules). This is something like a bunch of ping-pong balls endlessly bouncing around without the effects of gravity or any other influence. There is absolutely no direction or intention in the molecules themselves; the overall movement of perfume to the nostrils across the room is simply due to probabilities that some of the jiggling molecules will randomly end up there. An equal number of perfume molecules move by random jiggling in all other directions from the opened bottle. In fact, and this is really important, most of the perfume molecules that originally move out from the bottle randomly change course and head in different directions, including back toward the source bottle!

The significance of diffusion in biological systems is that there will always be an overall movement of molecules from an area of greater concentration (more molecules) towards an area of lower concentration (fewer molecules). For example, if oxygen molecules are more common outside a cell compared to inside a cell, then there will be an overall movement of oxygen into the cell. This is what happens when the human bloodstream drops off oxygen around our tissues. But why is there an *overall directed* movement of oxygen molecules into the cell? It is due to absolutely random, chance movements of oxygen particles. Diffusion is based on probabilities; more molecules will jiggle into the cell from the larger source outside, even while some molecules will move randomly in the opposite direction. There are uncountable examples in biology where the requirements of an organ or organism are met through the predictable, directional process of diffusion and, yet, it is entirely based on the random, non-directional movement of individual molecules.

There are important lessons in these illustrations. First, God's natural creation contains molecules that jiggle with kinetic energy and continually move and collide in random motion. This is a fundamental property of the physical world that God established. It is what he intended. Christians all proclaim the sovereignty of God and the Bible affirms that he actively sustains his creation.[2] Believers can be confident that God is in some manner responsible for the movement of all molecules in the universe. However, we must also concede that our perceptions are accurate when we describe molecular motion as completely random and non-directed. We are left with the conclusion that God's influence in the world is enacted while there are also natural events that appear to us to be determined by chance. I urge Christians to deeply ponder this reality even if our emotions or simple reasoning suggest something different. I will have much more to say on the topic of divine action in upcoming chapters.

A second lesson in our examination of diffusion is that the random, non-directed movement of molecules leads to an orderly, directional result upon which all living things depend. Our bodies rely on diffusion in innumerable ways (e.g., to move nutrient molecules from the intestines to the blood, to move neurotransmitter molecules across the synapses between neurons). This shows that Christians need to be careful when they reject chance events by attempting to equate them with a lack of direction or purpose.

I have been fascinated with how difficult it is for my students to accurately understand how diffusion works. The most common tendency is for individuals to explain that molecules "need" to spread out or that the particles only move in one direction from where molecules are most abundant to wherever there are fewer. Even some of my best students have to grapple with relaxing their assumptions about how things *should* work. At the heart of this struggle is an inherent philosophical sense in everybody (both Christian and non-Christians) that events in nature will be describable only on the bases of decisive, directional processes. This is a fundamental misunderstanding and an inaccurate description of some parts of the natural world. We are shown, instead, that randomness is an important aspect of phenomena that serve crucial functions in organisms.

Another example of chance events in nature is the scattering of leaves that drop from broadleaf trees during autumn. The leaves are released from individual trees in no precise sequence and they land on the ground in random array. Part of the remarkable beauty of these yellow, red and purple canvases scattered on the forest floor is how they end up arranged in wildly unpredictable patterns, something that can't be duplicated by human designs! Even so, the majority of the leaves will be broken down by soil fungi and bacteria and the nutrients will be recycled back into the trees for the

2. Col 1:17, Heb 1:3.

orderly reconstruction of new leaves the following spring. Again, random elements do not mean lack of order or ultimate purpose.

At the subatomic level, physicists have shown it is impossible to predict the location of electrons orbiting around an atom's nucleus because electrons move with so much random, unpredictable motion. In fact, a key principle of quantum mechanics is that chance events are a fundamental aspect of subatomic particles. However, regardless of the unpredictability of electrons in a carbon atom, we can count on the relative stability of the whole atom within a precisely construct protein molecule in a human brain.

The point, of course, is that an insistence that God's purposeful creation cannot involve random processes misapplies the concept of God's sovereignty and overly restricts what it means to say his creation is intended. In contrast, the natural world demonstrates that chance mechanisms can be an important part of meaningful, functional phenomena. Christians should not reject evolution because elements of chance are included in its descriptions. I will later explain the integration of random genetic events within the directional course of evolution.

God is More

Something I really enjoy in teaching biology is surprising students with how much in the natural world is different than our expectations based on human-centered experiences. For example, one obstacle in teaching bird identification is getting individuals to accept that baby songbirds are not smaller than adults. In numerous species, a very young, two-week old fledgling is already close to the size of its parents when it leaves the nest. This is something like every two-year old human toddler equaling the size and weight of its mother (yikes!). Although this seems "wrong" to our senses, nature instructs us that our initial perceptions are a narrow view of reality.

For Christians, a detailed study of the natural world reveals a God who has ordained and apparently takes pleasure in more than we can imagine. A variety of animal, algae and fungi reproduce asexually by casting off swimming cells or wind-borne spores that grow into duplicate individuals. This is the equivalent of me releasing single cells from my fingertips that settle and develop into exact Gary Fugle copies (each one like an identical twin). How strange this seems to us and, yet, how wonderfully fascinating! Most of us know that bees are attracted to colorful flowers, but we are unaware that once these insects land they are guided to nectar by intricately etched patterns emitted in ultraviolet light ranges perceptible to bees but not to humans. These UV patterns, that are part of God's surprising creation, were unknown to humans until scientific techniques revealed them. My study of biology has continually opened my eyes and taught me that God is increasingly more complicated, expansive and different than I once perceived or expected.

The key aim of this chapter is to encourage Christians to be careful about limitations we may bring regarding what God would or would not have done in the past in the derivation of living things. There are sound reasons for freeing one's thinking from very short time frames and idealized pictures of the past. It is especially important not to react against what science tells us about evolution (or anything else) on the grounds that it somehow doesn't make sense to us or doesn't appeal to us. Some of the most frequent expressions I hear about evolution are seemingly rhetorical questions such as, "How could evolution possibly lead to the intricate human brain?" or "How could evolution give rise to the perfect structure of a bird wing?" The unrecognized underlying argument is that a person's difficulty comprehending how it could happen is sufficient to discount it. I suggest it is far more important that biologists have been astonished at how extensively observations in the natural world describe an evolutionary past. Biological evolution is more complicated and, for some, emotionally less satisfying than special creation events, but I assume that what we all desire is the truth, not simplicity or pre-established notions of how God should act. As creationists, all Christians assert the doctrine that God brought the physical world and the diversity of life into existence, but we can be less certain about the specific methods he used to do so.

In fact, I suggest that in accepting evolutionary understandings, Christians are actually allowed to better appreciate the God we worship. By explaining the natural world masterpiece of the Creator, evolutionary biology reveals untold depth, complexity, elegance, and grandness. We see that God is even larger and richer than we previously perceived under the limits of simple creation accounts. I believe we rightly honor God by investigating and understanding the wonders of evolution conveyed in his created world.

The Bible teaches us to wrestle with who God is. This is a good thing. Job (the righteous man "in whom God delights") never wavers in his belief of God's authority and power, but he struggles with questions about God's actions. "I would present my case before him and fill my mouth with my arguments" (Job 23:4). When we believe we have a particular section of the Bible completely figured out, God will invariably be ready to teach us more about it. Consider the Pharisees who searched the Scriptures and yet were unable to see Jesus about whom it testified (Acts 10:43). We are always best off being open to changes in our perspective about a God whose ways are not wholly knowable. I encourage Christians to consider evolutionary ideas without anxiety, not as a compromise, a sad surrender or a dilution of our faith, but as part of our ever-expanding, glorious view of the God we worship.

6

Foundational Views in Christian Faith

CONSIDER THE ONE PERSON in your life you admire and respect above all others and then imagine your emotions if someone attacked his or her personal character and tried to alter your opinion. This can give us a sense of how many evangelicals feel when their deeply valued perceptions about the Christian church are challenged. This chapter addresses the concern that any attempt to merge evolutionary ideas with Christian faith is a degradation of foundational church beliefs that the faithful must resist. I have seen Christians hold tenaciously to specific creationist views with this expressed motivation. In fact, it is in this context that I have had my integrity attacked and the validity of my faith openly questioned. Arguments for evolution based on scientific evidence will have little influence on someone who already embraces such a defensive posture. Once again, I suggest this is a response largely based on misconceptions and misinformation. It not only dissuades Christians from honest consideration of evolutionary theory, it also present a false picture of orthodox Christianity.

People are generally surprised to hear that a six-day, young-earth interpretation of creation has not been the predominant view through the history of the Christian church. In fact, many early church theologians rejected the interpretation of Genesis creation "days" as exclusively six consecutive 24-hour periods and many did not perceive a creation event occurring only several thousands years ago.[1] This includes Augustine in the fifth century (generally considered a foundational figure in the development

---

1. Pennock, *Tower*; Ross, *Creation*.

of solid Christian doctrine). Augustine wrote, "What kind of days these were it is extremely difficult, or perhaps impossible, for us to conceive"[2] and "But at least we know it is different from the ordinary day with which we are familiar."[3] Some writers argued that creation did not take place in real time since time was created along with other things that exist. Others noted that a 24-hour earth day was not in existence until the sun appeared on the fourth "day," so there was no logical basis for measuring days prior to this by the typical sun-based standard.[4] In sum, a good number of early church leaders acknowledged that the first chapter of Genesis was a challenge to understand and interpret. It is especially significant that these individuals made these conclusions without pressure from modern science to see things this way. This history shows that the verses of Genesis 1 have never simply or obviously implied six 24-hour days or a very young earth.

There was also no united interpretation of the Genesis time scale through subsequent centuries of the church. Church scholars continued to hold varying views and an open, tolerant approach to the issue. For example, Thomas Aquinas in his *Summa Theologica*, written in the thirteenth century, discussed arguments for both six literal days of creation and for interpreting the days in some other manner. He wrote, "In order to be impartial, we must meet the arguments of either side."[5]

The Protestant Reformation brought a move toward a more definitive acceptance of the six-day, young-earth view. Martin Luther had a strong sense that the early biblical passages should be read as literal accounts. He wrote that Moses (historically thought to be the author of Genesis) "spoke in a literal sense, not allegorically or figuratively."[6] Luther therefore believed that the creation story occurred over six normal earth days. It is no surprise that Luther's views would have an impact on Christian churches that arose from the Reformation movement. However, it is quite relevant in this context to point out that Luther also used literal interpretations of Genesis to oppose proposals of his day that the earth orbited around a centrally positioned sun.[7] Like most of his contemporaries, he believed that the earth was

---

2. Augustine, *City of God*. Book XI, Chapter 6.

3. Augustine, *Literal Meaning*, Book V, Chapter 2.

4. "Now who is there, pray, possessed of understanding, that will regard the statement as appropriate that the first day, and the second, and the third, in which also both evening and morning are mentioned, existed without sun, and moon, and stars—the first day even without a sky? Origin, *De Principiis*. Book IV, 1:16.

5. Aquinas, *Summa*, First Part, Question 74.

6. Pelikan, *Luther's Works*, 5.

7. "It is likely that the bodies of the stars, like that of the sun, are round, and that they are fastened to the firmament like globes of fire . . . Scripture simply says that the

the centerpiece of creation while the sun, moon, and stars were attached to a dome-like firmament surrounding the earth. This was verifiable to Luther by a straight reading of Genesis 1, as well as by the everyday experience that the heavenly bodies transverse across the sky.

Although a six-day creationist position was common during the Reformation, spontaneous creationists have generally misrepresented the strength of commitment to the view. This is especially true with regard to prominent Reformation leader John Calvin. In his *Commentary on Genesis* (1554), Calvin refers to scientific pursuits with great admiration and repeatedly points out that Moses wrote so he would be understood by the common person rather than to impart scientific accuracy. Calvin proclaimed, "He who would learn astronomy, and other recondite arts, let him go elsewhere."[8] He even discussed how Moses was inaccurate in scientific detail by indicating that the moon produces its own light (i.e., Genesis 1:16 states that "God made two great lights"). Calvin noted that the moon doesn't actually shine on its own although it appeared to by the observations of people in Moses' day.[9] Calvin's perspective is expressed well in the following excerpt (although the translation from Latin takes some effort to follow):

> Moses wrote in a popular style things which without instruction, all ordinary persons, endued with common sense, are able to understand; but astronomers investigate with great labour whatever the sagacity of the human mind can comprehend. Nevertheless, this study is not to be reprobated, nor this science condemned, because some frantic persons are wont boldly to reject whatever is unknown to them. For astronomy is not only pleasant, but also very useful to be known: it cannot be denied that this art unfolds the admirable wisdom of God. Wherefore, as ingenious men are to be honoured who have expended useful labour on this subject, so they who have leisure and capacity ought not to neglect this kind of exercise. Nor did Moses truly wish to withdraw us from this pursuit in omitting such things as are peculiar to the art; but because he was ordained a teacher as well of the unlearned and rude as of the learned, he could not otherwise fulfill his office than by descending to this greater method of instruction. Had he spoken of things generally

---

moon, the sun, and the stars were placed in the firmament of the heaven, below and above which heaven are the waters . . . " Pelikan, *Luther's Works*, 42, 43.

8. Calvin, *Commentary*, Vol. 1, 1:6.

9. Calvin, *Commentary*, Vol. 1, 1:15.

unknown, the uneducated might have pleaded in excuse that such subjects were beyond their capacity.[10]

Calvin communicates here that the study of astronomy reveals the "admirable wisdom of God." He indicates that we should value scientific knowledge (it is "very useful to be known") and even honor those who pursue it. Calvin further warns against condemning scientific study or rejecting new scientific information as it is revealed ("because some frantic persons are wont boldly to reject whatever is unknown to them"). Lastly, Calvin explains that Moses was necessarily constrained from depicting scientific detail by his real purpose of reaching ordinary people with a communication about the character of God. In other words, the biblical writer intentionally did not provide adequate information for making strong scientific conclusions.

Examining Calvin further, when he pointed out in his writings that Genesis 1 refers to six days, he did so to refute an important contrasting notion of his time that creation occurred instantaneously. He never addressed an issue of 24-hour days compared to longer periods of time. Calvin's emphasis was a theological message, not a scientific one; God prepared his creation in a succession of time to demonstrate his loving care and provision for us.[11] Taken in total, Calvin's communications do not imply a man who was an avid young-earth creationist.

What's more, Calvin refers in his *Commentary* to "fixed stars" and the "earth at the center of the heavens," both common perceptions of the time, but now known to be false. This is important. If any spontaneous creationists insist on using John Calvin or Martin Luther to defend six 24-hour creation days, then by consistency they should be using them to argue for an earth-centered universe and stars fixed in position or attached to a dome-like firmament!

There is a tendency to assume that statements with scientific implications made by theologians centuries ago would be repeated by the same individuals today in the context of modern scientific understandings. This assumption is not valid. We will never know where Calvin might stand if he were alive today amidst creation-evolution discussions, but I can certainly make the case based on his respect for scientific evidence and his focus on Scripture's theological message, that he would be receptive to mainstream scientific discoveries and to evolutionary creation. Leading post-Darwin

10. Calvin, *Commentary*, Vol. 1, 1:16.

11. "For the same reason, the world was created, not in an instant, but in six days. The order of creation described, showing that Adam was not created until God had, with infinite goodness made ample provision for him." Calvin, *Institutes*, Book I, 14:2.

theologians, such as B.B. Warfield at Princeton Theological Seminary, have interpreted Calvin in the same way.[12]

Tolerance for different interpretations of the Genesis 1 days continued through the Reformation. It is noteworthy that the highly regarded Westminster Confession of Faith of 1646 did not specifically designate 24-hour creation periods and this has been widely understood as purposely allowing for differing views.[13] It wasn't until Bishop Ussher's infamous chronology in the middle 1600s and its subsequent inclusion in printings of the King James Bible that many Christians outwardly conceptualized a very young earth. This was a significant historical event as a particular individual's interpretation of the creation timeline became equated with the authority of the Bible. Still, the position was never universally received. By the middle of the 19th century, many liberal *and* conservative Christian apologists were particularly influenced by the findings of geology and readily conceded that the Bible allowed for an ancient earth and a flood of local extent.[14]

Even evolutionary concepts were not generally met with concern in the worldwide church after the initial introduction of Darwin's ideas in 1859. Many conservative theologians actually welcomed the idea as both scripturally sound and beneficial because it implied God maintained continued involvement in nature through evolutionary development rather than acting only in the past through special creation.[15] These theologians were evolutionary creationists. It is an unfortunate aspect of history that among Darwin's friends and supporters, the anti-religious biologist Thomas Huxley gained more notoriety and press than did botanist Asa Gray. Gray and other Christian scientists and theologians admirably defended the compatibility of evolution and Christian orthodoxy, but as is still true today, controversy garners most of the media's attention. Theistic evolution (evolutionary creation) became such a prevalent view by the end of the 19th century that some writers suggested earlier conflicts over evolution and Christian theology were over.[16]

However, it would be an inaccurate portrayal to suggest that there was not important opposition to an ancient earth and evolutionary proposals. One of the most highly celebrated and influential preachers of the past two centuries, Charles H. Spurgeon, had this to say in 1877:

12. Warfield, "Calvin's Doctrine."

13. Chapell, *President's Goals.*

14. Numbers, *Creationists*, 3–36.

15. Livingston, *Forgotten Defenders.*

16. Ramm, *Christian View*, 284.

> We are invited, brethren, most earnestly to go away from the old-fashioned belief of our forefathers because of the supposed discoveries of science. What is science? The method by which man tries to conceal his ignorance . . . Forsooth, you and I are to take our Bibles and shape and mould our belief according to the ever-shifting teachings of so-called scientific men. What folly is this![17]

Spurgeon should be rightly honored among evangelicals for composing some of the most insightful and inspirational sermons ever delivered to Christians. And yet, this quotation breaks my heart and I would hope that other devoted followers of Christ and his Word will readily recognize the improper fallacies it contains.

Spurgeon's unabashed rejection of scientific information and derogatory declaration regarding those who would pursue it foreshadows the remarkable and unsettling approach of many modern-day theologians and young-earth creationists. Spurgeon would have us isolate ourselves from scientific revelations and rely solely on our study of biblical passages and past beliefs to declare what humans can discover to be true about God's natural creation. This is precisely what occurs in many conservative seminaries and churches of our time in which well-educated individuals trained in biblical studies and theology determine, in isolation from scientists, what evangelical believers should accept regarding earth history. But rejecting science and using the Bible as the sole source of knowledge about all things is not church tradition. While Spurgeon appeals to the understandings of important forefathers, his approach actually strays very far from the teachings and wisdom of Augustine, Aquinas, Calvin, and others.

I believe Christians are better instructed by the inspirational words of conservative theologian B.B. Warfield. Warfield is well known among evangelicals as a champion of biblical inerrancy.[18] He was also committed to the understanding that God reveals himself through his creation as well as through the Bible. He wrote in 1893:

> We must not, then, as Christians, assume an attitude of antagonism toward the truths of reason, or the truths of philosophy, or the truths of science, or the truths of history, or the truths of criticism. As children of light, we must be careful to keep ourselves open to every ray of light . . . Let us, then, cultivate an attitude of courage as over against the investigations of the day. None should be more zealous in them than we. None should be

17. Spurgeon, *All-Round Ministry*, 97.
18. Hodge and Warfield, "Inspiration."

more quick to discern the truth in every field, more hospitable
to receive it, more loyal to follow it, whithersoever it leads.[19]

Warfield accepted that the days of Genesis were long ages. He was also open
to evolutionary progression as a mode of divine action. Warfield was careful
in his writings to describe his understanding that God may have guided
evolution providentially through natural laws. At the same time, he rejected unacceptable naturalistic explanations of agnostic scientists.[20] Warfield was an evolutionary creationist along with a good number of biblical
scholars of his time.

More familiar opposition to evolution arose out of the Fundamentalist
movement in the early part of the twentieth century.[21] Ironically, the organized Fundamentalist cause did not begin with a rejection of evolution as
a mechanism of creation and almost all participants were very comfortable
with an old age of the earth. Instead, Fundamentalism arose in response to
liberal "modernist" challenges to core aspects of Christian doctrine, such as
biblical inerrancy, the reality of the virgin birth, the resurrection of Christ,
and biblical miracles. A formative document at the time was a four-volume
set of 90 articles titled *The Fundamentals*. It was written by leading conserva-
tives to confirm the central principles of Christian faith. Only around a fifth
of the articles made any reference to evolution, there was little mention of
24-hour creation days, and although there was some opposition to evolution,
the essays contained no general instruction to resist evolutionary teachings.[22]

However, the Fundamentalist movement galvanized many conservative Christians and rallied them to protect core beliefs. In time and especially among the general population, opposition to evolutionary thinking
became part of the package. This was the context for the infamous Scopes
evolution trial of 1925. It may be good to remember that, in contrast to
many people's view of the creation-evolution conflict today, William Jen-
nings Bryan, the Fundamentalist champion and prosecuting creationist
attorney was an old-earth creationist like most of his contemporaries. Evolution was generally downplayed or excluded from many American schools
through the first half of the twentieth century. This was at least partly due to
sparse coverage in textbooks, which publishers wanted to sell to local school
districts. Meanwhile, evolutionary theory was becoming fully established
as the major unifying principle within biological disciplines at colleges and
universities throughout the world.

19. Warfield, "Incarnate Truth."
20. Noll and Livingston, *B.B. Warfield*.
21. Giberson, *Saving Darwin*, 59–61; Numbers, *Creationists*, 20–36.
22. Giberson, *Saving Darwin*, 60.

Finally, it was not until the 1960s in America that young-earth creation gained its current popularity. It is generally agreed that the modern phenomenon surged with the publication of *The Genesis Flood* by John Whitcomb and Henry Morris in 1961 and the advent of several well-organized young-earth creationist organizations. There is now an enormous commitment to a six-day, young-earth position among evangelical Christians in this country. This perception is perpetuated by erroneous reports from young-earth advocates that "consensus of the church through all these centuries is that the earth and the universe are very young, only several thousand years old."[23] The remarkable shift in the meaning of "creationism" from a theologically more orthodox old earth/local flood interpretation to the current young-earth understandings is one of the most notable sociological events in America over the last half century.[24]

It is probably important to note that the rise in popularity of young-earth creation had been a distinctly American phenomenon that was not mirrored in the Christian church in other parts of the world until well into the 1980s. As a person who came of age during the current creation-evolution debates, I once assumed that this state of affairs had been the case continually for hundreds of years. Instead, I was surprised when I discovered that what we experience today is actually a relatively recent phenomenon and a unique controversy in the history of the Christian church. Importantly, there has also been strong concurrent opposition and concern from within evangelical denominations about the influence of modern "creationists." For example, Mark Noll, a conservative Christian historian, has referred to young-earth creation as "a fatally flawed hermeneutic of the sort no responsible Christian teacher in the history of the Church has ever endorsed."[25]

The point of this chapter is that the historical Christian church has not strongly defended any particularly creationist position and that evolution has not always been viewed as conflicting with faith. No specific formulation is "foundational" or the "correct" Christian position. Instead, details regarding when and how God created have been and continue to be topics for open investigation and discussion. The only universal Christian doctrine of creation through history has been that God created the heavens and the earth and all that is in them.

Even further, real Christian orthodoxy as demonstrated by leading church theologians has been to wrestle with and reevaluate specific interpretations of biblical passages, not hold fast to single readings. The

23. Mohler, "Universe," 5.

24. Numbers, *Creationists*.

25. Noll, "Scandal," 28.

great thinkers in history were humble to the fact that their conclusions were always subservient to the real truth known only by God. Many past theologians accepted a six-day creation event because they were given no particular reason to believe otherwise. At the same time, read the writings of Augustine, Aquinas, and Calvin and you will see that they consistently valued and celebrated intellectual pursuit of truth revealed in God's natural creation. The strongest Christian tradition has been to actively formulate biblical understandings that coincide with natural revelations. Given our expanding knowledge of the natural world, we should grant that the best interpretations expressed in the past are not necessarily the best interpretations now. In this way, when believers consider the encompassing view of evolutionary creation we are actually following the inspirational lead of great Christian thinkers and foundational theologians in church history.

PART III

The Collision of Ideas

IN THIS SECTION, I turn to prominent issues that have arisen where the application of Christian faith and science intersect. Two focal areas are the practice of science in professional settings and the delivery of science education, particularly in public classrooms. Although I only addressed complicating assumptions and emotions of Christians in the previous chapters, I will now also include challenges that face practicing scientists and science teachers. As I hope you will see, it is through adjustments to commonly held beliefs and approaches of both the Christian church *and* the scientific community that we may appreciate why and how creation and evolution can be positively and effectively brought together.

7

The Face of Science

It has been fascinating over the years to hear comments from fellow evangelical Christians when they discover I am a college biology professor. Many will say they are glad for my presence as a Christian in such a godless environment. Others will comment on how good it is to have someone there to speak against the exclusive voice of evolution. I smile and find a careful way to point out that I don't hold all the views they might imagine and I politely suggest they may have a false impression of many of my colleagues. Even so, I have come to recognize that the sentiments of these believers are not entirely off the mark.

Science Defined

One of the most important considerations in creation-evolution discussions is establishing an accurate and workable description of what natural science is. In a single sentence, here it is: natural science exclusively studies the properties of matter and energy in the natural world, and its hypotheses, predictions, tests, and conclusions are based on empirical evidence pertaining to these properties. It is intriguing to note that the historical origin of science as a way of knowing ("natural theology") sprang from a *religious* understanding that the Creator imparts and sustains his natural creation with consistent parameters and processes. In other words, Christians believe science is made possible because of who God is and they celebrate the Creator when they engage in science. The philosophical conviction that the world operates through discoverable properties is called "naturalism." Since the methods of natural science focus exclusively on these features, this approach to learning and knowledge is called "*methodological* naturalism." A significant aspect of this description is that science specifically does not study or explain God or any other supernatural elements because these are beyond and are not defined by the regular parameters of the natural world.

Science, *as a process*, also must proceed with the assumption that God or other agents do not intervene because that would be an alteration in the fundamental regularity in natural properties upon which the discipline depends. A scientist can explain how a ball tumbles down a hillside by describing the gravitational force that pulls it downward and the friction with the ground that causes the ball to roll during its descent. But we are immediately outside the realm of science if we say that a supernatural agent might unpredictably and by unknown means cause the ball to reverse direction and slide up the slope.

It follows that when science is properly conducted, it does not inherently provide theistic information, ethical direction, or value judgments. The latter, instead, arise from a person's worldview, a total picture or set of reasoned beliefs about our existence (see chapter 2). In religious formulations, one assumes the existence of God (or gods) and a supernatural or metaphysical realm that is beyond the confines of the natural world. In contrast, a purely naturalistic worldview assumes that matter and energy and their interactions constitute all there is to reality. The latter assumption is "*philosophical* naturalism." A vital distinction exists between this philosophical position (a belief that the natural world is all that exists) and the methodical naturalism of science (an approach for studying the natural world). Philosophical naturalism amounts to atheism while methodical naturalism does not.

In these descriptions, one can see that there is no direct link between science and particular metaphysical worldviews. In other words, both a deeply religious person and an atheist can practice pure methods of science with equal success and no conflict. One of the great attributes of scientific endeavors is that they fundamentally operate without any influence from our religious and cultural differences and at the same time do not have a particular goal of informing these aspects of our lives. A key communication of this chapter is that the distinction between science, as a method of studying the natural world, and the different worldviews to which individuals adhere must be consciously and actively maintained. Unfortunately, this is precisely the context where many or most creation-evolution conflicts arise. The following discussion will venture into situations where individuals, from Christians to atheists, and from scientists to nonscientists, knowingly or unknowingly confound this distinction and contribute directly to debates based on miscommunications and confused issues.

In the Defense of Science

Misunderstanding about science and worldviews has been played out most notably in struggles over science education in public schools. Most people are well aware that mainstream scientists oppose religious-based creation science and intelligent design hypotheses and these conceptualizations have been consistently excluded from science classrooms by courtroom decisions. Christians often interpret this as a godless scientific establishment rejecting religion. This is not, in principle, an accurate assessment. The response mistakenly equates defense of evolutionary theory and rejection of God-centered hypotheses within science as a rejection of God and religion. These are not the same thing. Instead, mainstream scientists are vigorously defending the realm of science as one that is limited to the study of natural mechanisms that have operated and continue to operate in the world. While God is excluded from the methods of the discipline, this does not mean there has been a wholesale rejection of spiritual or religious realities. Strong defense of what science *is* does not constitute any position whatsoever about the existence, importance, or relevance of God and metaphysical truths. *To exclude God from the methods of science does not exclude God from life.* Christian believers must be careful to differentiate between these two things.

As creation-centered hypotheses within science have been denied in courtrooms and at colleges over the last many years, spontaneous creationists continue to characterize it as rejection of religion. One of the strange but illustrative aspects of this battle is that creation science and intelligent design hypotheses are almost always promoted as *nonreligious* formulations to assure their scientific credibility. But when a vast community of both non-Christian *and* Christian scientists rejects them as improper *science,* many spontaneous creationists tell us it is due to *religious* persecution and an atheist agenda. This is not fair or logical. The argument starts with an acknowledgement of the distinctive nature of science but ends with the confused communication that a person's religious views really are part of and should be permitted to redefine the scientific process. In contrast, scientists must be able to disallow creation science and intelligent design proposals because they see them as invalid erosions of the fundamental character of science without being charged with judgments pertaining to God and religion (see the next two chapters for more complete discussions of these issues). The crucial point is that the quality of all scientific ideas must be evaluated separately from faith. I wholeheartedly join my scientific colleagues in defending this purity of the scientific method.

However, even if we can agree on these clear distinctions, defending the character of science is hardly the only matter in the everyday world of science and religion. A closer look reveals that there are real reasons why Christians are wary or even fearful of scientists and the scientific method, particularly in the realm of evolutionary theory.

The Silence of Religion within the Scientific Community

One element of concern for Christians about science in general is that there is a culture within the scientific community that does not accommodate religious beliefs. The lack of religious recognition is true even though many scientists are practicing Christians, Jews, Muslims, and followers of other faiths. The culture partly stems from the fact that the practice of science itself, which specifically answers questions without invoking God, inclines scientists to exclude God from the professional setting. It is also true that the majority of scientists have agnostic or atheistic leanings. With this mixture of factors, there seems to be an unspoken understanding that spiritual issues should be diplomatically separated from the everyday practice of the scientific profession.

Even beyond this, the relative absence of religion from scientific culture results from subtle to open hostility and suppression by nonreligious individuals. It is difficult to gauge the precise nature and extent of this pressure across various institutions, but there is little doubt that it exists. Kenneth Miller, a prominent academic defender of evolution and a Roman Catholic believer, describes what he calls a "fabric of disbelief" throughout the academic establishment based on the assumption that deep religious faith is something individuals abandon as they become more intellectually informed.[1] Some biologists believe (and would say) that they are outwardly tolerant of religious colleagues and students but subtly communicate their disapproval and promote an assumed superiority of atheistic thinking. Others are more blatantly condescending. To be clear, I am referring to behaviors that go beyond the defense of science as a naturalistic methodology to communications that disregard or deride religious views.

Given this backdrop, it is not surprising that many Christians distrust the motives of the scientific community. In particular, they have reason to suspect the scientific basis for evolution and they understandably embrace spontaneous creationist positions presented by individuals with

1. Miller, *Darwin's God*, 184–85.

philosophical worldviews more like their own. Published statements from major science organizations, such as the National Science Teachers Association and the National Academy of Science, acknowledge that spiritual conclusions are reached outside the realm of practicing science: "Because science is limited to explaining natural phenomena through the use of empirical evidence, it cannot provide religious or ultimate explanations" (NSTA) and "science can neither prove nor disprove religion" (NAS).[2] If the scientific community truly understands these statements, it follows that its participants should soundly reject even subtle religious suppression and more explicitly accept the varying spiritual lives of colleagues. Scientists should strongly protect the purity of science and reject misguided religious notions that impinge upon it. But, to be balanced in that message, they should also decidedly respect and protect the religious perspectives of colleagues and students that are beyond the natural explanations of scientific endeavors. There is no doubt that a meaningful transformation in this regard would be beneficial in improving the relationship between scientists and nonscientists in American society. Ultimately, scientists who understand the repercussions of what I am communicating must influence the members of the scientific community who prefer to justify or ignore insensitive and suppressive behaviors that exist.

Clarifications

I'd like to be obvious to Christians about what I am referring to here as religious suppression. Christians should be respected and not receive condescending comments that their beliefs are "unenlightened" or "irrational" (or anything of a similar nature). Even so, a believer is not being persecuted when he or she is corrected for a poor understanding of science. My experience is that scientists, whether atheist or otherwise, are honorable in terms of evaluating academic and scientific work of Christian students and researchers. This is because it is not difficult to separate the value of work, based on its scientific quality, from the faith of the person who produces it. For example, if a student writes a class essay on the invalidity of evolution when the assignment was to discuss how biologists understand a speciation event might occur, they will rightly be penalized for their poor understanding of the scientific material, not their Christian faith. This is *not* persecution or restriction of free speech. Similarly, if a scientist claims there is no or limited evidence for biological evolution and utilizes God as a scientific explanation

2. NSTA, *Teaching of Evolution*; NAS, *Science, Evolution and Creationism*, 54.

for observations in nature, he or she will be rightly challenged by colleagues for the poor science independent of their personal spiritual beliefs.

I am aware that the degree of intolerance and pervasiveness of atheistic thinking varies widely from place to place. I have had rewarding discussions with colleagues about religion and my faith at every institution at which I have studied and taught. Many scientists who insist on an atheist position in the workplace are best described as agnostics. If nothing else, a "there may be a God" formulation is most consistent with a scientist's trained tentative approach to knowledge. If pressed, many agree they are not certain we live in a godless existence. I have found that once fellow biologists learn I am a capable scientist, most are open to hear about my Christian worldview. What is certainly clear is agnostic/atheist scientists are most concerned about protecting the integrity of their scientific discipline and they adamantly reject forms of Christian faith that insist God-centered hypotheses within science are a necessary aspect of belief.

On the other hand, it would be naive for me or other scientific professionals to claim that Christian individuals within academia are not the object of various kinds of derision. Even though colleges and universities are generally liberal institutions that are well-known for fostering and tolerating a diversity of thought, traditional Christianity is often viewed as the least fashionable position, if not a disfavored one. Many scientists can learn to be more sensitive and inclusive with both students and colleagues. I recently attended a conference with a number of college biology educators. At one point, a couple of individuals poked fun at creationist notions and the religious individuals who accept them. Although I agreed with the scientific basis of their critiques, I wondered what the response would have been if I had clarified, "I celebrate the power and majesty of God in the natural world." I am fairly certain the speakers would have been more careful in their wording if they knew that I (and likely others in the room) have a Christian worldview. But why should I have to be confrontational to correct this bias? Biologists should always discuss evolution with the assumption that listeners include both evolutionary creationists (i.e., religious individuals) and atheist evolutionists.

Insensitivity by scientists may be subtle or perhaps even unintended. For example, a popular college text in the late 1990s included the phrase, "Darwin compiled enough support for his theory of descent with modification to convince most of the scientists of his day that organisms evolve without supernatural intervention." Of course, science cannot tell us that God had no involvement whatsoever in the evolutionary process. Encouragingly, later editions of the same text dropped this declaration and rather stated that Darwin "convinced many of the scientists of his day that organism do indeed evolve."[3] A heightened consideration for religious perspectives can be achieved.

There are also cases of more outward insensitivity. The National Association of Biology Teachers released a "Statement on Teaching Evolution" in 1995. The document stated, "The diversity of life on earth is the outcome of evolution: an unsupervised, impersonal, unpredictable, and natural process of temporal descent with genetic modification that is affected by natural selection, chance, historical contingencies, and changing environments."[4] The words "unsupervised" and "impersonal" have no value in a scientific definition of evolution and, instead, inappropriately indicate a metaphysical belief that God is not involved. Fortunately, these descriptions were removed from later versions of the publication. There was resistance to the improved modifications by atheistic scientists who would use their platform to assert a personal belief system[5], but we must remain hopeful that the previously quieter, but larger, voice of more sensitive and inclusive scientists will prevail in these situations.

The noteworthy point in this discussion is that antireligious behaviors of scientists can be just as significant a contributor to the creation-evolution divide as Christians who promote religion-based science. I am sure many scientists will first be inclined to object to this statement, but I believe that a careful examination for unperceived bias or elitism will reveal its validity. The fact remains that if scientists go beyond the communication and defense of science itself by surrounding it with atheistic expectations, then they are communicating just what Christians fear. This gives Christian believers good reason to make an evolution-atheism connection and this pushes them to choose nonscientific forms of creationism. It is common within the scientific community to pass judgment on Christian groups that won't listen to the sound reasoning of evolutionary theory, but scientists should be aware they are often conveying an atheistic worldview that interferes and makes evolution unacceptable. As I exhort Christians to accept the findings of evolutionary biology, I urge fellow scientists to more outwardly accommodate varied religious perspectives so that evolution can be positively studied and understood.

3. Campbell et al., *Biology*: 2nd ed., 262, 4th ed., 260.

4. NABT, *Statement*.

5. Giberson, *Saving Darwin*, 167–69.

Scientific Claims for Atheism

The most important reason many Christians are wary of science, and particularly biological evolution, is that some prominent scientists blatantly claim that it justifies and leads directly to atheism. This is going beyond suppressing religious expression within the scientific community to communicating that atheism is the correct position based upon science itself. When biologists draw such anti-theistic conclusions, they drift into a realm that is well beyond the discipline of science. It is also beyond their realm of authority as *scientists*. One of the poster figures in the atheist group has been Richard Dawkins, author of several best-selling books, who outwardly bashes anti-evolutionary challenges and theistic positions. Dawkins famously stated almost 20 years ago that "Darwin made it possible to be an intellectually fulfilled atheist"[6] and Dawkins' communications continue today. Pronouncements like this have occurred consistently over the last few decades from prominent scientists, such as Isaac Asimov, E.O. Wilson, and Carl Sagan, and more recently in a surge of "New Atheist" writings.[7]

These individuals have a right to their own opinions, but it is misleading and non-beneficial for them to express it as a strong conclusion from science. They have fueled the creation-evolution controversy by explicitly stating what many spontaneous creationists insist is true: that acceptance of biological evolution is ultimately equivalent to an atheistic worldview. The error, of course, is that no matter how much of the world can be explained scientifically, a conclusion that natural processes are all there is to existence is a philosophical choice based on faith. Atheistic authors have gone beyond scientific conclusions to make metaphysical declarations that flow from the philosophical naturalism they embrace as their encompassing worldview. There have been numerous rebuttals by both scientists and theologians.[8]

A particular challenge for Christians (and all individuals of faith) is atheistic conclusions reported in popular science media. For example, a recent article in *Discover* magazine summarized fascinating research on specific parts of the brain that are active when humans solve moral dilemmas. In an otherwise excellent piece, a lead researcher is quoted, "You have these gut reactions and they feel authoritative, like the voice of God or your conscience." Then the article's author continues, as if the researcher's quote goes on, "But these instincts are not commands from a higher power. They are

6. Dawkins, *Watchmaker*, 6; *God Delusion; Magic of Reality.*

7. Dennet, *Breaking the Spell;* Harris, *End of Faith;* Harris, *Moral Landscape;* Hitchens, *God is Not Great.*

8. Collins, *Belief;* D'Souza, *What's So Great;* Haught, *God;* Hitchens, *Rage;* MacGrath and McGrath, *Dawkins Delusion;* McGrath, *Why God.*

just emotions hardwired into the brain as we evolved."[9] There is no difficulty with the idea that neuron circuits in specific brain regions promote emotional, instinctive responses during moral decisions, but why make reference to God or claim these areas are "just" the result of evolution? Ethicists can easily argue that these brain regions evolved precisely as God intended. Christian readers are well aware of inappropriate atheistic insertions like this in science reporting.

Atheists Don't Win

It is certainly worth noting that atheism is only a small minority view in our country, as most polls demonstrate that around 90% of all Americans believe in God.[10] Atheistic pronouncements are also out of step with the majority of individuals within academic institutions. For example, a national survey of American college and university faculty across all academic disciplines found that 81% "consider themselves a spiritual person" and 69% responded that they "seek out opportunities to grow spiritually."[11] Even among practicing scientists, 40% acknowledged a committed religious view that they believe in a God who actively communicates with humanity (a "God to whom one may pray in expectation of receiving an answer").[12]

In contrast to atheist statements, many scientists combine a strong Christian worldview with a full acceptance of modern science. There have been excellent offerings by evolutionary creationists; including some very readable books that effectively speak to the evangelical Christian community.[13] I have previously mentioned the helpful work of the BioLogos Foundation. This organization's commitment to biblical Christian faith, as well as to the ability of modern science to reveal truth, is captured in this statement released after an initial conference of 47 organizing pastors, theologians, and scientists near the end of 2009:

> We affirm that the truths of Scripture and the truths of nature
> both have their origins in God, and that further exploration of

9. Ohlson, "Morality," 32.

10. e.g., *Newsweek* poll, March 30, 2007.

11. Higher Education Research, *Spirituality*.

12. Larson and Witham, "Scientists."

13. Alexander, *Creation or Evolution*; Collins, *Language of God*; Falk, *Coming to Peace*; Giberson and Collins, *Science and Faith*; Haarsma and Haarsma, *Origins*; Peters and Hewlett, *Can You Believe*.

all these truths can enrich our joyful and worshipful appreciation of the Creator's love, goodness, and grace.[14]

Other organizations include the American Scientific Affiliation, the Colos-
sian Forum, the Center for Theology and The Natural Sciences, and the
International Society for Science and Religion. These groups investigate
the intersection and reconciliation of science and religion and among their
members are many individuals who see no conflict between evolutionary
theory and Christian faith based on a high view of Scripture. There are also
evolutionary creationist websites scattered throughout the Internet.

The evolutionary creation message has not penetrated as deeply as it
might into the Christian community at large. This is due to a variety of factors. One reason is that many Christians are unaware of the well-developed
arguments for evolutionary creation by prominent theologians and scientists. Another is the intellectual challenge required for adult Christians to
adjust previously held convictions. Unfortunately, I also believe vocal and
powerful spontaneous creationists have, in many circles, effectively silenced
such a message and those who would favor it. Still, I am certain there is a
desire among many for an encouraging, integrated view of modern science
and biblical faith to be heard within the evangelical church and across our
society in general.

The fact that there are simultaneous arguments for atheism and Christian faith from different practicing scientists demonstrates that neither position is an apparent conclusion from the methods of science. We should
fully expect that both scientists and nonscientists will debate the validity
and value of different worldviews, and I would hope that we continue to
encourage respectful debate on the topic. On the other hand, I believe most
alert individuals can recognize the importance of distinguishing this kind
of debate from defenses of the purely naturalistic endeavors and discoveries
of science. Christian believers should at least be able to acknowledge that
atheists are not representative of all scientists.

The concerning issue for Christians in atheistic declarations is not the
discipline of science itself, but the individuals who argue we should reject
spirituality and God based on scientific information. It is not science or
its methods, but the choice of humans who prefer to explain and solve the
questions and challenges of life with God excluded. For some people there
is a tendency for scientific or technological knowledge to lead them away
from God and the value of spiritual truths. I suggest to Christians that we
should not be entirely surprised or alarmed (e.g., "the world through its
wisdom did not know him . . . ," 1 Cor 1:21). For individuals who do not

14. BioLogos, *Statement*.

have a strong spiritual perspective, this can be a reminder that mechanistic explanations may not be all there is to full wisdom. Whatever science may point out to us, there is always a bigger view of our existence, and the largest questions about meaning and purpose still remain. The most meaningful focus for Christians is to question the spiritual decisions of individuals. In other words, Christians may eagerly embrace exciting natural discoveries of modern science wherever they may lead, including biological evolution, because these discoveries are good science. Separately, we should have a reasoned and effective voice to argue that a Christian worldview is a desirable alternative to atheism. The Church has always known that the real battle is whether or not people turn their hearts toward God.

An Unfortunate Christian Counter: the Darwinist Agenda

The previous pages have described a picture of the scientific community that generally goes unrecognized or unacknowledged. Given the atheistic associations I have mentioned, it should be understandable that many Christians have had anti-science reactions. Sadly, the result has too often been inaccurate and misguided overreaction. For example, some spontaneous creationists imply that *all* evolutionary biologists are promoting atheism and are a threat to the Christian church and what it values.

The Institute for Creation Research is a group dedicated to young-earth creation. It opened a Museum of Creation and Earth History in San Diego during the late 1990s. Author Robert Pennock recounted his visit to this museum soon after it opened.[15] One of the exhibits on display at the time had two opposing panels contrasting the "fruits" of creationism and the "fruits" of evolutionism. The creationism panel contained:

True Christology	True Faith	True Family Life
True Evangelism	True Morality	True Education
True Missions	True Hope	True History
True Fellowship	True Americanism	True Science
True Gospel	True Government	

The "fruits" of evolutionism on the other panel were:

Communism	Racism	Abortion
Nazism	Pantheism	Euthanasia
Imperialism	Behaviorism	Chauvinism
Monopolism	Materialism	Infanticide

15. Pennock, *Tower*, 315.

Humanism	Promiscuity	Homosexuality
Atheism	Pornography	Bestiality
Scientism	Genocide	
Slavery	Drug Culture	

Clearly the first view is an association of creationism with God, positive values, and ultimate truth while the second connects evolution with a godless existence, evil, and immorality. Similar associations of evils with evolution are communicated through other young-earth creationist groups, including Answers in Genesis.[16] From this vantage point, the creation-evolution issue is not just a determination of correct scientific formulations, but a battle for a proper life perspective that will lead us to good ethics, solid morality, and correct politics. "Creation" and "evolution" are stark titles for what is seen as the grand opposition of two critically different worldviews.

An acceptance of this false dichotomy, in either the extreme form illustrated here, or in more subtle variations expressed by some intelligent design advocates, is a major force behind resistance to evolutionary biology in America. The mainstream scientific community is characterized as the "Darwinian establishment" that controls scientific thought and has an "agenda" to eliminate God and religion from our schools and public life. I have been told on many occasions that evolutionary thinking is merely a conspiracy trumped up by individuals who must justify their atheistic beliefs. Henry Morris, the late ICR leader, has stated, "The fact is that evolutionists believe in evolution because they *want* to. It is their desire at all costs to explain the origin of everything without a Creator" (Morris' emphasis).[17] We are told by intelligent design authors that even evolutionists who are Christians are promoting atheism because the methodological naturalism of science is essentially equivalent to metaphysical naturalism. According to Philip Johnson, a founding leader of the ID movement, "The Darwinian theory of evolution is not merely or primarily a scientific theory . . . It is, first and foremost, a creation story for the culture . . . The story tells us that we were created by blind and purposeless material processes rather than by a purposeful Creator."[18] These notions have been picked up and regularly announced by many Christians (most of whom have no scientific affinities and limited scientific training), including Christian organizations, radio stations, conservative political commentators and, most significantly, church leaders.

16. Ham, "The Evil Fruit"; Bartz, "The Religion."
17. Morris, "Evolution is Religion."
18. Johnson, "Shouting Heresy," 23.

Spontaneous creationists who make these statements have mis-framed the conflict and misdirected their attack at all evolutionary biologists and the scientific community in general. By sweepingly equating evolutionary science with atheism and spontaneous creation with a belief in a Creator-God, a scientific topic is reduced to a spiritual debate. This is unquestionably inaccurate and nonproductive. It forces a false dichotomy of two unattractive choices and ignores the most reasonable option that evolutionary biology and a belief in God are completely agreeable. The real battle is *not* between evolution and creation, but over the worldview people choose to embrace. The latter is something behind which all Christians can unite, a choice to live a life without God or a life with the redeeming grace and love of a transforming Savior.

It is quite remarkable how frequently spontaneous creationists state these charges against evolutionary biologists, and I plead with Christians to consider the offensiveness of such declarations. Biologists as a whole are not setting out to exclude God from our culture, and they do not have a secret pact against Christian faith. Evolutionary biology is a major discipline at almost every college and university in the world because it is based on supportable ideas. To realistically imagine how it could be retained with a weak and suspect framework of evidences is the grandest of conspiracy theories. Geologists and biologists in the 1800s came to the conclusions that the earth is very old and that evolution has occurred because they could no longer deny what the natural world reveals. It is not because they were searching for scientific information to back a nonreligious position. Charles Darwin waited many years before releasing *The Origin of Species* precisely because he knew well and was concerned about the theological implications of his ideas and findings. As a biologist, I defend the validity of evolutionary ideas and assert that the scientific position is the product of honest investigation and thought. Christians are wrong to attack the science of evolutionary biology in an attempt to counter that with which they really disagree, a dismissal of God and a Christian worldview.

One of the odd consequences of claiming that evolution is tantamount to atheism is that it confronts a large body of fellow Christians like me who have sincerely come to a position of evolutionary creation. We accept methodological naturalism as the rule in science and conclude that evolutionary theory is well-supported, but we also believe strongly in a biblical doctrine of creation. And yet, we are charged with succumbing to indoctrination and inaccurate judgments on evolution and are told that we are part of an anti-church conspiracy.

Sadly, it is my experience that it takes as much or more determination to be an evolutionary creationist within the evangelical church than a

Christian within the scientific community. I have received more derision from fellow believers for my evolutionary views than from scientific colleagues for my Christian worldview. I see people roll their eyes or shrug their shoulders and tell me that I just don't understand or that I've been influenced by propaganda from the scientific establishment. It still astonishes me that some individuals will so easily dismiss my position on something that pertains to my lifelong profession. The remarkable part is that this often comes from Christians who have done very little study and have a poor understanding of mainstream evolutionary proposals and evidence. Typically, they are emboldened by an exclusive exposure to young-earth creationists and intelligent design publications. All I can say is that I have never had a scientist so forcibly tell me what kind of spiritual worldview I should have, but I've repeatedly had fellow Christians tell me how I ought to do science. There is something terribly wrong here.

The most unfortunate thing is that many non-Christian biologists who might be entirely open to honest discussion about Jesus and spiritual matters are justifiably turned away from the Christian church by attacks from spontaneous creationists who give them no respect or acceptance. It is one thing when fellow believers tell me how to do science, but it is altogether another when they inform my non-Christian colleagues that they are indoctrinated and are wrong about something held with high confidence in the scientific community. I caution followers of Christ to think more carefully before voicing charges based on an inaccurate association of evolution with atheism.

In summary, the evangelical Christian community should not claim that biologists are rejecting God and religion when they defend the purity of science as methodological naturalism. Even though practicing scientists possess a diversity of religious beliefs, almost all of them deem that attempts to inject God as a scientific explanation are entirely inappropriate. Also, evangelicals should not insist that there is an unavoidable connection between evolution and atheism and make individuals choose between two fictitious choices of (1) atheism and evolution versus (2) God and spontaneous creationism. Instead, the dialogue should be opened to embrace evolutionary creation so that biblical creation and the findings of modern science are both upheld. As evolutionary creationists are allowed to make a greater impact on the debate, non-Christian individuals are freed to enjoy the wonders revealed by science *and* to consider the claims of Jesus.

The Future of Science Education

I finish this chapter by offering a new perspective for scientists as they communicate their discoveries to the general public, especially within the science classroom. The fundamental role of science education is to delineate the significant worth of scientific methodology as a "way of knowing." Our lives have been incredibly enriched as investigators continually push for scientific explanations of phenomena in the natural world. An appreciation for this approach, and the use of critical thinking and problem solving in applying it, are a vital aspect of modern education. But, for most individuals, science is a method that addresses only certain kinds of truth. As prominent evolutionary biologist Francisco Ayala has stated, "A scientific view of the world is hopelessly incomplete."[19] What should science educators do with religious perspectives?

Those of us in education are often asked to use holistic or interdisciplinary approaches. This is because it has been shown that the best learning happens when we place new information into the context of our life experiences. In nationwide surveys of over 100,000 freshman college and university students, 77% answered that they "believe in God," 69% noted that they "pray," and 58% responded that it is "very important" that they "integrate spirituality into their lives."[20] This is the real world in which scientists teach biological evolution. Some topics in the sciences necessarily intersect with and challenge the religious faith and ethical framework of students. Even though this is obvious, science educators generally do not acknowledge the need for students to construct a coherent worldview that incorporates all their acquired knowledge.

In addition, most scientists misunderstand how most of the general population inherently integrates scientific and religious information. The scientific community almost always views the relationship between science and religion as two entirely separate spheres of knowledge. Stephen J. Gould, one of the most well-known evolutionists (and one of my academic heroes), argued that the two disciplines are endeavors that speak to two distinctly different aspects of our existence; science is the examination of the natural world while religion (or theology) is the study of God and spirituality, and by extension, moral and ethical values. Gould famously referred to the two areas of study as "non-overlapping magisteria."[21] The view can be visualized

19. Ayala, *Darwin*, 102.

20. Astin et al., *Cultivating*, 83.

21. Gould, "Nonoverlapping Magisteria," 22; This viewpoint is comparable to the "Independence" category in Ian Barbour's typology for relating science and theology (*Religion and Science*, 77–105).

as two independent circles (see Fig. 7.1a). Gould, a self-proclaimed Jewish agnostic, promoted mutual respect between practitioners of science and religion and acknowledged that both areas contribute to wisdom. He even wrote, "I believe, with all my heart, in a respectful, even loving concordat between our magisteria." I am sure that many of my nonreligious scientific colleagues would voice equally gracious sentiments. Interestingly, many or most scientists with religious faith also endorse or go along with this perception of separate domains.[22] Even so, the notion of non-overlapping magisteria, or any viewpoint that divides science and religion into separated realms, has notable implications and unrecognized consequences.

First, the perceived compartmentalization of science and religion draws a very distinct line of separation. This leads to an easy (and convenient) justification that science classrooms are for the exclusively study of scientific findings while religion should be covered in different courses of philosophy, religious studies, or other disciplines. In other words, under non-overlapping magisteria scientists can rightly isolate themselves from spiritual discussions. We reason that once students pass through the doorway into a biology class they will hear only about natural phenomena and we can insist on total independence of science from other forms of thought. I absolutely agree that a science curriculum should focus almost entirely on the details of natural mechanisms, but the current norm is to so intently communicate and maintain this focus that the spiritual aspects of students are in effect ignored or inadvertently devalued. Most religious individuals go along with this approach, but many have a real sense that they have been shortchanged. The result is a real or perceived barrier between the disciplines of science and the religious values of students and, by extension, of the general population.

A contrasting Christian view of the relationship between religious understanding and science is reflected in Fig. 7.1b. Science is understood as the focused study of the natural world, but unlike the prevailing model among many scientists, God is seen as sovereign over both the spiritual and natural realms. Christians don't separate God exclusively into a supernatural role. Instead, we believe that God created both the natural and metaphysical realms and that he is intimately involved in all aspects of existence. Christians believe that the ongoing mechanisms of the natural world, upon which science depends, are themselves dependent on God. In addition, believers expect that an omnipresent God interacts with these processes to guide the course of events. With this inherent approach to life, Christians necessarily look to incorporate the findings of science into one coherent view that includes God, not compartmentalize scientific information and spiritual understandings into separate spheres of knowledge.

22. Ayala, *Darwin*, 91.

(a) Science and religion as non-overlapping ways of knowing

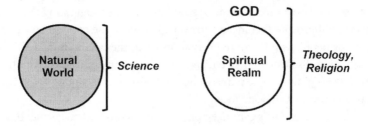

(b) Christian view of science and religion

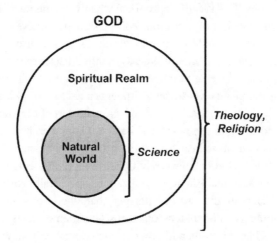

Figure 7.1 — Different views of the relationship between science and religion. (a) Science and religion study two separate spheres of our existence; science studies the workings of the natural world while religion (or theology) studies God, the spiritual realm and philosophical issues of ethics and morality. This is the view communicated most commonly by scientists. (b) Science studies the natural world, but this is a subset of the entire realm over which God has involvement and which religion (or theology) studies. This is a Christian view.

This may not seem like a big deal to scientists but the seemingly accommodating approach of non-overlapping magisteria, and corresponding arguments for complete exclusion of religious references from science classrooms, are inherently unsatisfying and alienating for students who are religious. If evolution is understood to be the best scientific explanation for the diversity of life, there is an immediate question for Christians as to where and how God has acted or is involved in the process. Science educators often proceed with a false sense that religious questions have been safely set aside into a separate magisterium, but we are ignoring that this is not the mind-set of many or most students. I am not suggesting that scientists should provide theological answers. However, the fact remains that all religious people will be more likely to listen to the details and significance of scientific discoveries if scientists show familiarity and sensitivity to the real way religious people perceive human existence, including the spiritual issues with which they are faced.

Science educators can make practical decisions and take small but significant actions to address the science and religion topic in a different way. I propose that most introductory science courses and textbooks should purposefully contain a succinct section on the relationship of science, religion, and philosophy. This is fully appropriate when we remember the real-world context that three out of every four college students believe in God. In particular, instructors should be able to communicate the distinctive nature of scientific endeavors and science as a way of thinking (see the beginning of this chapter), but also be able to impart that science can be fully integrated into encompassing spiritual worldviews. Students need to hear a clear rationale of why God-centered proposals are not a legitimate part of a science curriculum (further addressed in the next two chapters). However, it is also important for religious students to hear from their instructors that science cannot either prove there is a God or conversely disprove God. Students should be told that most professional science organizations specifically communicate this position.[23] They should also be informed that a good number of practicing scientists embrace different religious faiths. Christian students can be directed to resources like this book and other references that will allow them to pursue questions more deeply. The essential suggestion is that students, and all nonscientists, can be proactively shown that their spiritual perspectives are safe and can be integrated with scientific findings. Without these communications, the face of science and scientists *is* atheistic and we appear to promote philosophical naturalism, not just methodical naturalism.

Currently, these messages are not regularly conveyed. Some science educators are atheists who consciously or inadvertently condone their own worldview by avoiding the science-religion distinction or by making inappropriate atheist announcements. Instead, they should explicitly acknowledge that their position is a matter of philosophical choice rather than an obvious conclusion based on science. It would not matter even if all science

23. e.g., "The benefits (of scientific understandings) can be lost if science education is undermined either by those who claim that science is the only source of knowledge or by those who claim that scientific knowledge is only legitimate if it conforms to a particular set of religious doctrines." AAAS, *Study Guide*, 3.

colleagues agreed with an atheistic position (there is no indication that they do); teachers must not abuse their authority. Some educators, regardless of their own spiritual views, simply want to do science and hope that they can isolate themselves from the religious aspects of our culture. But this ignores holistic instruction that stimulates students to listen and learn, and it increases the likelihood that students will disengage when evolutionary topics arise. There are resources available for educators to consider how to effectively discuss worldviews within public classrooms.[24]

Some scientists may rightly object that it is difficult to acknowledge religious perspectives while Christian groups simultaneously pressure science teachers, school boards, and legislators to include inappropriate creation science and intelligent design concepts in science classrooms. Unfortunately, this is valid in the sense that some Christians could (or likely would) misinterpret religious recognition as acceptance of these proposals. Even so, I fully believe it is possible to defend against the unscientific nature of these propositions while also showing respect for and acknowledgment of religious perspectives in general.

Some may call attention to the fact that science educators (as well as research scientists) are not equipped to take on presentations or discussions of science and spirituality. I would respond that this is precisely the point. Why do we throw high school and college science instructors into a biology classroom with no preparation in this area? Currently, these individuals typically receive instruction only within the details of their favored scientific discipline. Science teachers (at least) can easily be exposed to training in the philosophy of science and science-religion issues. Courses of this type are increasingly common on college campuses. It is also reasonable to suggest that biology instructors become familiar with some of the evolutionary creation resources mentioned in this book. Requirements of this nature in teacher preparation, and an assumption that instructors will deliver a small science and religion presentation as part of their science curriculum, would result in a significant beneficial shift in science education. Although I do not have the naive belief that this would solve the creation-evolution wars, I do believe it is appropriate and would help dissolve tension and misunderstandings that now exist.

I fully anticipate that many within the scientific establishment will object because of the already existing challenge of inappropriate religious conclusions within natural science. As mentioned above, this is a legitimate concern. But does this mean science educators should refrain from

---

24. AAAS, *Dialogue on Science, Ethics and Religion*; Rusbult, *Worldview Education*; Teaching About Religion.org.

working toward the most effective instruction? The idea is not to discuss creation science or intelligent design as scientific concepts, rather it is about respect for religion. We are collectively in a sad place if the real objection is a simple and entrenched fear of any religious acknowledgement within the context of a science classroom. What I am describing here is a new paradigm in which religious sensitivity is a simple, normal, and expected aspect of science education.

8

It's Only a Theory

MOST CREATION-EVOLUTION DISCUSSIONS TODAY have their roots in some fascinating history from the 1970s and 1980s, a history that can teach us many things. During this time, organized efforts were developed to establish the claims of young-earth creation as scientific conceptualizations. "Creation science" or "scientific creationism" rose to the point of a social and political phenomenon. This perspective remains strongly embedded among evangelical Christians. The creation science model is a series of statements about the natural world that flow directly from a six-day, young-earth reading of Genesis. Most of the elements are phrased in a purposeful manner to exclude reference to God or the Bible. For example, the model includes: (1) the earth is only several thousand years old, (2) the earth's geologic features are the result of rapid, catastrophic processes during a worldwide flood, and (3) all living kinds of organisms have remained fixed within describable limits of genetic variation. Creation scientists argue that these statements can be treated as scientific hypotheses and we may, in turn, collect information from the natural world to test their validity. Following this line of reasoning, creation science was and still is promoted as a genuine scientific alternative to evolutionary theory. Proponents in the '70s and '80s lobbied for "equal time" or "balanced treatment" in science classrooms.

The push for scientific creationism had considerable influence on local school boards and state legislatures over a short period. By the early 1980s, legislative bills that promoted creation science in public classrooms had been introduced in 20 different states.[1] The rebuttal by our nation's court systems and the scientific community was unambiguous and firm. Several state court rulings and finally a U.S. Supreme Court decision in 1987 banned scientific creationism from science curricula. Not surprisingly, young-earth creationists characterize this as a biased, secular conclusion. But what is the real basis for the refutation?

First, creation science is rejected because it is a religious construct in which the "hypotheses" are derived from one particular Bible interpretation rather than from observations in the natural world. The Supreme Court ruling noted that the intent of creation science legislation was unmistakably to restructure the science curriculum to conform to a particular religious viewpoint. Young-earth proponents can hardly argue with this assessment since they outwardly assert that we should begin with a biblical presupposition that God created the universe over six days only several thousands of years ago. The courts pointed out that the normal starting point for scientists is making observations in nature. Hypotheses based on the workings of natural processes are then formulated in an attempt to explain what is observed. Scientists do not pursue evidence for predictions based on the actions of God (even if the predictions are phrased as if God has been excluded). On this basis alone, the repeated judgment is that scientific creationism is not simply a set of scientific ideas, but a religious doctrine.

In contrast, evolutionary theory is not generated from any inherent metaphysical worldview (atheist, agnostic, Christian, or otherwise). In fact, no young-earth creationist can clarify how the extensive and complex explanations of evolutionary biology could have developed *solely* from a philosophical position. In addition, no creation scientist has ever argued that someone limited to observations in nature could independently come up with the creation science model and its predictions. This is incredibly important!

The second reason creation science has been rejected is there is very little evidence to support what the model predicts. Many young-earth creation leaders have agreed that current scientific information is against them, but they hold out hopes that new, as yet undiscovered, evidence will vindicate their position.[2] Darrell Falk, from the BioLogos Foundation, reports three meetings with steadfast young-earth creationists (one choked with emotion and tears) who faithfully hold to their commitments even though they acknowledged that the natural world does not provide support.[3] I feel

1. Pennock, *Tower*, 4.

2. Nelson and Reynolds, "Young Earth Creationism"; Schadewald, "Conference on Creationism"; Wood, "The Truth."

3. Falk, "One Hundred Fifty Years."

for these fellow evangelical Christians and I respect their devotion to what they believe God has called them to defend. Even so, one can hardly argue that theirs is a *scientific* position. In the meantime, most of the purported evidence for young-earth creation has been extensively refuted and there are excellent resources that explain their weaknesses.[4] In other words, regardless of any religious connotations, creation science predictions have not stood up well in explaining the natural world. I suggest there are very profound reasons why this should matter a lot to all of us.

Meaningful Science

Fair-minded and compassionate individuals from a variety of philosophical and religious viewpoints have suggested that evolutionary ideas and creation science should be taught alongside one another in public schools. Since evolution is only a theory, it seems like young-earth creation should be comfortably offered as an alternative. Why can't biologists be fair about this and compromise? Aren't scientists the ones who are causing the controversy by being obstinate and stubborn? It's easy to see why people are drawn to this line of thinking; it seems so reasonable and conciliatory. Even so, there is something eerily wrong here.

The phrase "only a theory" is an oxymoron to scientists. A theory in science is a well-supported collection of ideas and explanations; its concepts have been tested and confirmed by numerous observations. The theory of gravity in physical science or cell theory in biology are examples. "Theory" does not mean, as implied by "*only* a theory," a speculation or a guess. The latter is what scientists call a hypothesis, but even this is an "educated guess" based on observations of the natural phenomena under study. In our common language, people like to use the word, "theory," to mean just about any idea or proposal someone has, but this is not at all what scientists mean.

A distinctive character of the scientific process is that it promotes skepticism and purposely avoids the notion that the final word has been spoken on any subject. Ideas are always subject to new testing, analyzed from new angles, and open to modification. Currently supported and accepted explanations are viewed as our best conceptualization for the present. If science is to remain dynamic and forward-moving, there must be future investigations and an anticipation of new findings. Scientists like to say an idea is "well-supported" rather than "proven." In principle, if a person gets to a point where an idea is simply accepted and new research is deemed unnecessary, then the practitioner is no longer doing science.

4. Isaak, *Counter-Creationism*; NAS, *Science*; Scott, *Evolution*; TalkOrigins, "Index."

However, it is improper to interpret this to mean that all ideas in science are just working hypotheses and that all ideas have equal merit. Useful scientific proposals allow us to make predictions about what we should see in the natural world and we should then be able to make further observations and/or conduct manipulative experiments to see if the predictions are correct. Science educators make an effort to teach that the best hypotheses are "testable" (we have to be able to gather information to distinguish if the idea is valid) and "falsifiable" (gathered information should not just be consistent with the hypothesis, it should allow us to clearly reject or accept the idea). If predictions from a hypothesis hold up, scientists continue to pursue the notion with increased vigor, but if predictions are repeatedly not met, the proposed scientific explanation is dropped as incorrect. In this way, new discoveries correct inaccuracies and refine current understandings. Through the process, certain concepts in science are valued for their ability to explain what is observed in the natural world while other proposals fade away for their inability to do the same. Scientists understand that this is how science progresses and, over time, they almost always agree which proposals are well-supported and which are removed from the table.

When we talk about evolution, it is helpful to note that the scientific approach works well for describing and understanding *past* events and processes just as it works for explaining currently observable natural mechanisms. Some spontaneous creationists promote the false notion that science can only address phenomena that are shown through experiments in the present. This is a self-imposed limitation that implies all historical conclusions are simply opinions based on preselected assumptions, but the notion is logically false and more rhetoric than substance.

All historical sciences work with discoveries that inform us about past events. When I discover a vacant nest and distinctive egg fragments of Cassin's Vireos during the cold winter months, I can confidently conclude that a pair of these birds nested in my yard during the previous spring. In the same way, paleontologists may uncover a distinctive dinosaur fossil that helps to construct a picture of past life on earth. In line with all scientists, evolutionary biologists make predictions about the nature of historical evidence that is yet to be discovered. This new information is used to test hypotheses and it has the power to support, falsify, or improve current explanations.

A Theory Emerges

If a given set of explanations is repeatedly supported by an enlarging body of confirming observations, scientists put increasing confidence in their validity and start referring to this set of ideas as a "theory." For the biological community as a whole, the general concept of evolution and some ideas about the mechanisms of evolutionary change (e.g., natural selection) have reached this level of support. An obvious implication is that the general body of ideas is so well-supported that it is unlikely to be overturned. This is true even though certain details of the theory will be modified or expanded as new information is obtained.

As new ideas have come along in evolutionary biology, they have caused a refinement in previous understanding. Punctuated equilibrium, for example, was a successful addition to our thinking regarding the pacing of evolutionary change. Previous conceptualizations emphasized gradual alterations over time, but there are actually very good, formerly unrecognized, reasons to expect relatively rapid modifications in some circumstances. Also, the discovery of major control genes in animal development, called Hox genes, has opened new avenues for understanding the path of evolution in animals. We now understand that only tiny changes in these control genes can lead to very big differences in an animal's adult form. However, this work to modify specific formulations about the processes of change has not altered the overall support for evolution or very much of what we already understand about mechanisms. The concepts of evolutionary change over large time periods, major ancestral relationships among organisms, and our current understanding of things such as modern genetics, reproduction, and natural selection collectively comprise what biologists call evolutionary theory.

We can now return to the suggestion that evolutionary biology and creation science should be taught together in science classrooms. My purpose in the previous paragraphs has been to illustrate precisely how convinced mainstream scientists are by the evidence for a long evolutionary history on earth (see chapters 11–15). This counters the common and too often repeated understanding that evolution is merely a speculation about the derivation of different life forms. If the latter were an accurate perception, the intent of the claim, "evolution is only a theory," would be wholly justified. But this is not close to an accurate characterization of evolutionary thinking in biology.

The scientific community values evolutionary theory as an incredibly useful, unifying, and guiding principle in biology. We understand organisms more fully as we see them within a framework of ancestral connections and it explains innumerable aspects of their biology. Far from being a peripheral speculation, biologists use evolutionary thinking every day as a fundamental aspect of their discipline. It is often difficult for nonscientists to understand how incredibly exciting and satisfying this is for biologists.

No scientists from the distant past could have ever imagined we would have a conceptual structure that explains both variations in DNA molecules and the structure of entire biological communities. There is no other framework in biology that is so remarkable and comprehensive. Under the creation science model, we are asked to believe this is all concocted nonsense.

As a scientist, given what I have seen and studied, I feel I would have to sacrifice both the methods of science and logic to not accept the general phenomenon of evolutionary change over millions of years. Of course, from what I have experienced and studied in my spiritual life, I would have to sacrifice the core of who I am to not acknowledge there is a Creator-God. This is why I am an evolutionary creationist.

Science Moving Forward

There are certainly times during the progression of science that two different proposals may appear to explain the same observed natural phenomena. This is the proper context for two competing ideas to be simultaneously discussed in scientific pursuits and public dialogue. However, natural science also must be an advancing activity of refinement and improvement during which certain ideas become verified and others lose favor.

For example, there was a debate in the past about how the incredibly important process of photosynthesis gives off oxygen gas. Scientists had deciphered that plants use light energy to convert simple molecules of carbon dioxide (CO_2) and water (H_2O) into complex sugars ($C_6H_{12}O_6$) and oxygen (O_2). The prevailing understanding for a long time was that the oxygen gas came directly from the precursor carbon dioxide molecules. On the other hand, some researchers made the intriguing suggestion that the O_2 byproduct somehow came about from the oxygen atoms (O) that start out in water. These two alternate descriptions were both in play until more research was conducted. Eventually, investigations falsified the first idea and confirmed the latter; we now know that water molecules are split apart and two oxygen atoms from water are combined together to make the oxygen we breathe. Nobody continues to debate this conclusion.

A more familiar example of scientific advance comes from competing proposals during the seventeenth century of an earth-centered versus a sun-centered existence. Although there was a time period in which both ideas could be defended, accumulated scientific observations led to rejection of the earth-centered model and confirmation that the earth orbits the sun as part of a planetary system. There is now no serious opposition to this common understanding.

The point is this: modern scientists view the creation science model in the same light as other past proposals that don't do a good job of explaining what is observed in nature. I join fellow scientists in their frustration that perpetuation of creation science does not allow real science to move forward. Not only is creation science unable to explain a myriad of observations in the natural world, it does not even allow us to agree on many basic understandings. Imagine researchers who want to discuss the process of photosynthesis but are confronted by a small vocal minority that continue to insist that oxygen gas comes directly from carbon dioxide molecules. Consider a professional astronomer at a national conference who offers detailed investigations about the orbiting moons of Neptune but is challenged to defend the basic principles of planetary motion upon which his research depends. Evolutionary biologists have advanced so far into the subtleties of evolutionary phenomena that they are able to explain how closely related species are distributed into different habitats, how hemoglobin molecules have evolved in complexity through time, and how developmental control genes have generated different animal forms. And yet, we are challenged to defend the fundamental notion that evolution has occurred at all.

The crucial message is that creation science and evolutionary theory are not competing proposals that attempt to explain commonly agreed-upon observations. Instead, they are completely disparate notions with no middle ground. But there is no real reason why we can't conclusively establish whether the earth is 10,000 years old or billions of years old. Likewise, we really ought to be able to settle whether sedimentary deposits around the world are the result of a single global flood or slow accumulations over very long periods of time. With creation science and evolutionary theory both on the table, it is like we are arguing whether photosynthesis occurs at all, not about details within the process. All the while, there is one evident and legitimate way to determine what is true about the natural world (the world Christians agree was created by God): pursue detailed scientific investigations of natural evidence and let those findings uncover and lead to that truth, whatever it may be.

A Closer Look

Of course I can hear the loud objections of young earth creationists who continue to argue that their proposals are a legitimate contender. It is not possible to repeat analyses of creation science evidence provided in detail elsewhere, but it is helpful to examine the character of a few examples. First, creation scientists have described a variety of dating techniques that indicates a younger earth than that proposed by astronomers and geologists. The dating methods include examining continent erosion rates, measurements of helium addition to the earth's atmosphere, rates of stalactite formation, rates of salt deposition in the earth's oceans, decay of the earth's magnetic field, disintegration rates of comets, and many more. However, difficulties have been demonstrated in the logic and results for every one of these approaches, and scientists, to be honest, have easily dismissed them as unsatisfactory challenges to the very long time frames understood by the scientific community.[5] For example, Hugh Ross, an evangelical Christian and old-earth creationist with a Ph.D. in astronomy, reports that he analyzed *eighty* claims of scientific evidence for a young earth and found that every one was faulty, particularly due to strained underlying assumptions, inaccurate data, or misapplication of principles.[6]

What first appears to be substantial evidence for a young earth, based on the sheer number of examples, turns into a large collection of doubtful arguments that do more to bring the position into question rather than support it. Of particular importance in this discussion, creation scientists have largely ignored refutations of their dating techniques and these "evidences" continue to be communicated to Christians who will listen. Mainstream scientists recognize that there is no normal progression of science where unverified hypotheses are removed from consideration. This phenomenon is repeated with other discredited reports that continue to appear in young-earth creation literature. There have been very few events like this in the history of science: A vocal group of individuals has separated itself from the scientific community to continue with its proposals despite the fact that the vast majority of trained scientists have rejected what the group presents as supportive research.

A second category of evidence is observations that short-term local catastrophes can swiftly lay down sedimentary deposits that appear similar to those that modern scientists claim were deposited over much longer periods. This supports the creation science suggestion that a worldwide flood formed the earth's geological features. Major layering of volcanic ash and mud after the 1980 Mt. Saint Helens volcanic eruptions is a widely used illustration. The weakness here is that it is an unjustified jump to conclude that several localized large-scale sedimentation events today somehow means almost all the enormous quantity of sedimentation around the earth occurred rapidly in a *single* global flood event in the past. From a simplified

5. Matson, "How Good"; Thompson, "Is the Earth Young?"; TalkOrigins, "Age of the Earth."

6. Ross, *Creation and Time*, 103–108.

view of earth sediments, we can say that an enormous global flood *could* have rapidly laid down deposits all over the world, but the deposits also *could* have accumulated slowly over many thousands or millions of years.

This is an example of evidence that does not distinguish among alternatives (i.e., it doesn't falsify either option). Sedimentation by short-term catastrophes is loosely consistent with creation science proposals and this gives encouragement to supporters. However, scientists recognize the weak value of the information; it does nothing to falsify our understanding of sedimentation over geological time. Most important, modern scientists have other good reasons to conclude long-term sedimentation occurred. In particular, plant and animal fossils are distributed in sequences through sediment layers in chronological patterns that repeatedly match evolutionary predictions. Creation scientists have offered only strained speculations to explain how fossil sequences could develop by a large-scale flooding event (discussed in later chapters).

Finally, a central tenet of the creation science model is that organisms remain fixed within the limits of biblical "kinds." This notion is primarily supported by the repeated claim that there are "systematic gaps between fossil types" (i.e., there are no transitional forms). Oddly, the existence of large numbers of convincing transitional fossil forms (with more added on a regular basis) is fully accepted and never debated in the general biological community. Such a wide disparity in conclusions obviously has little to do with the actual physical evidence and is instead an impasse produced by established presuppositions.

For creation scientists, it is not possible that fossil remains will have intermediate traits that indicate major evolutionary transformations. The scientific community has witnessed that each presented transitional fossil is immediately rejected no matter what characteristics it possesses. But transitional fossils are obvious to evolutionary biologists and each new discovery fascinates us as it fills in the history of past life. Mainstream scientists continue to investigate natural mechanisms and analyze discoveries at their face value to further elucidate the details of an evolutionary past that has already been widely supported. The claim that there are no transitional forms asks us to deny major aspects of scientific progression over the last two centuries to return to a notion that was removed from discussion many years ago. Modern biologists are really left with no choice but to ignore proponents of scientific creationism and move on with evolutionary studies. Please note that this is not a case of blindly denying arguments of young-earth creation.

In summary, there are significant reasons for the scientific community to feel strongly about the exclusion of creation science. What value are our methods of science if unsupported proposals are kept in the mix only because they are favored and pushed by a vocal few? The issue is much bigger than scientists defending the specifics of evolutionary theory. Instead, mainstream scientists are resisting a challenge to the integrity of science itself. We cannot continue to include proposals that have not stood the test of natural evidence, no matter how strongly some people demand it. The latter inappropriately suggests that scientific conclusions are merely the result of a social struggle rather than based on data from the natural world.

Creation Science Today

One might reason that rejection of scientific creationism by our nation's court system and detailed refutation by scientists would have caused the conceptualization to fade in popularity. That is not what has happened. Creation science is still strongly advanced with great influence by young-earth creation organizations. They insist that the scientific community and our secular society are blinded because they exclude the authority of the Bible and will not allow the proper presupposition of young-earth creation to guide interpretations.

Creation science understandings are still widely embraced by the general population and it dominates the mind-set of many evangelical Christian churches. As indicators of the latter, young-earth creation materials are often the only or most prominent offering in Christian bookstores and the creation science framework is almost exclusively taught through widespread Christian home school curricula. Representatives from young-earth organizations visit churches and encourage believers to take hold of and defend this "proper" biblical view. Answers in Genesis, ICR, and other organizations offer extensive Internet websites and publish large numbers of articles, books, and glossy media for adults and children. AiG also operates an elaborate Creation Museum in Kentucky. Many evangelicals flock to the message and understandably conclude that the sheer enormity of written words and materials indicates validity in the young-earth position.

Meanwhile, almost all biologists proceed with the knowledge that the arguments for scientific creationism have been soundly invalidated to the point of having no credibility at all within the general scientific community. Practicing scientists understandably are not interested in endless arguments about a predetermined scenario that has not been supported by natural evidence. Most Christians are unaware that someone well-trained in evolutionary biology can take the creation science materials that circulate in isolation among evangelical communities and demonstrate that they are riddled with errors. Unfortunately, I can attest to the fact that these include false information, invalid rationale and, especially, misrepresentations of both evolutionary concepts and evolutionary biologists. I cannot tell you how many times an eager Christian friend or acquaintance has given me creation science literature to help correct my evolutionary thinking. In every case it has not been difficult to point out errors in the seemingly persuasive rhetoric and scientific arguments of the authors. For fellow Christians who are primarily familiar only with young-earth creation communications, I offer a compassionate yet pointed challenge: please consider these materials with open-eyed skepticism.

What Can We Conclude?

The full presentation of this chapter allows me to put the public confrontation between young-earth creation and evolution in a more clarifying light. I have emphasized that the only way to determine scientific legitimacy is to dive into the details of natural observations and data and let these things disclose what is true. In many areas of knowledge, from political science and economics to philosophical arguments, there may be multiple well-reasoned viewpoints that simultaneously remain in discussion to form our collective body of wisdom. The existence of different political parties demonstrates this point. But scientific pursuits are distinctly different in a very important way! In each specific aspect of scientific investigation we generally understand that there is inherently *only one correct empirical description*. In other words, truth in the natural world is a singular reality, an unchanging single thing that is waiting to be uncovered and described. Natural scientists have been revealing and describing various aspects of this reality for hundreds of years. Although there will always be more to unravel, our scientific knowledge has been built on past advances and our understanding has become increasingly intricate and detailed. With this backdrop, we should not be lulled into discussions as if we really don't know enough to say anything definitive about the history of life on earth. Some young-earth creationists make the sweeping claim that there is no empirical evidence for evolution. This is simply untrue. At some point, we must do the hard work of analyzing natural evidence and allow it to inform us.

However, this is not where the public dialogue on creation-evolution spends its time. Instead, we hear about people's opinions, their underlying biases and motivations, and their religious perspectives. We are incorrectly allowed to challenge the *character* of individuals who voice views opposing our own. Science is frequently characterized among Christians as an ever-changing endeavor with temporary conclusions that will simply be replaced by something new tomorrow. Popular evangelical pastor and author, Matt Chandler, offers the following, "I should lay my cards on the table and admit I'm a bit agnostic when it comes to science. Scientists change their minds quite often. Their theories evolve more than they say animals do."[7] Accepting a description like this allows Christians to dismiss unwelcomed scientific conclusions, but I urge fellow believers to understand that it is both inaccurate and inadequate. If scientific conclusions are really so elusive, then none of us should be relying on the knowledge and skills of doctors, believing that NASA scientists can send remotely-controlled robots to the distant surface of Mars, or trusting any historical descriptions of the Roman Empire. We can't really have it both ways and say we trust science in some areas and not in others. The truth is that scientists are fully capable of concluding many things with a great deal of certainty. This includes confidant assertions about an evolutionary past.

As the general public falsely perceives that science can't distinguish among alternatives, emotional appeals become far more important than scientific observations. In fact, most people come to understand that the creation-evolution topic is primarily our *choice* about what we want to *believe* is true. We hear and accept the phrases, "I believe in special creation" or "I believe in evolution." As we continue to think this way, the grand discourse over creation and evolution becomes competing advertising campaigns for the hearts and minds of individuals. But scientists are not advocates for different personal choices or campaigners who persuade others to believe in one of several viable options.

The notion that individuals may choose what they want to believe in science misrepresents what is supposed to be conclusions based on what is confirmed and falsified in nature. This downplays or ignores a wealth of natural observations. It is a largely accurate statement to say that young-earth creationists believe in special creation. On the other hand, practicing natural scientists do not accept evolution because they "believe" in the theory. It would not even occur to most of them to think of it this way. Instead, they acknowledge, in an unremarkable fashion, that evolution is extremely well-supported by empirical evidence and they embrace it because it is a comprehensive and useful scientific concept.

There is a crucially important principle at the heart of this chapter's communication: the scientific community's conclusion about the deep validity of evolutionary theory is accurate even if the majority of lay people in our country embrace young-earth creation, or if legislative bodies try to dictate that it be taught in science classrooms. Popular opinion or legislative

7. Chandler, *Explicit Gospel*, 92.

action cannot establish scientific validity, even though a part of us feels like that seems fair. There is a very strange and strong tension in our society over this issue. The general public desires to influence the conclusions biologists are allowed to make within their own disciplines based on nonscientific criteria. But the best scientific explanation can only be determined by the methods of science! It is not like determining a population's favorite soft drink, style of music, or presidential candidate based on personal preferences. This challenge is incredibly alarming to both Christian and non-Christian scientists who are concerned about scientific literacy in our country. They understand that the soundness of evolutionary theory compared to alternative proposals cannot be decided by popular demand or strength of conviction. Scientists are not being elitist; they are standing up for quality and reliability within their own field of study.

I do not believe that Christians in general, or evangelical Christian communities in particular, have fully grasped the situation I have described here. I must ask fellow believers to carefully consider if they are in fact doing the right thing as they tell each other that something is true about the history of the earth that is clearly understood to have little merit with experts in scientific disciplines who study such things. If large parts of the evangelical church continue to view creation science and evolutionary biology as a matter of choice (what we want to believe), there will never be real healing or progress. As I have done throughout previous chapters, I plead with believers for openness to evolutionary thinking that is harmonized with a biblical Creator-God, if not for the sake of an encompassing, accurate view of the world in which we live, then for the purpose of reaching others with the Christian gospel we value above all else. If change is to occur in many church environments, it will be up to individual Christians to acknowledge current conditions and object to one-sided promotion of young-earth creation. I enthusiastically anticipate a future in which fellow evangelicals actively engage in dialogue about how evolutionary biology can be most correctly incorporated with deep, abiding faith.

9

Design in Nature

FOR ALL CHRISTIANS, THE delicate structure of a bird feather and the complexity of a majestic oak reveal a wondrous God. We believe that all that is around us is here by the intent of a sovereign Creator. This often leads to the argument that the intricate form of living things is too astonishing to be the result of evolutionary processes. As odd as it may seem to some, we need to ask if the latter proclamation goes hand-in-hand with the previous assertions.

A new challenge to evolutionary biology emerged during the early 1990s. "Intelligent design" (ID) was first championed by a group of individuals associated with the Discovery Institute in Seattle, Washington, including Philip Johnson, William Dembski, Michael Behe, and Stephen Meyer. The core of this approach is to argue that there are complexities in biology that evolutionists cannot explain and that these features must have arisen by some means other than natural processes. ID proponents point to traits in living things that they suggest demonstrate "irreducible complexity" or high levels of "complex specified information." These include a number of biochemical and cellular features, intricate structures like the vertebrate eye and phenomena like the relatively rapid appearance of fossils types during the Cambrian geological period. If evolutionary processes cannot satisfactorily explain these things, then we may logically conclude they are the result of "intelligent design."

The ID movement may be most fully understood by distinguishing two separate aspects of its history. First, proponents argue that an intelligent designer can be utilized as a *scientific explanation* for observed natural phenomena. The key question here is whether the ID rationale should be included as an acceptable approach in scientific research and education. This is familiar territory where we need to think through the goals and consequences of scientific endeavors.

The second aspect of ID has been something quite different; it is an organized *cultural and political movement* to counter the dominance of materialistic philosophies in human society. The original group at the Discovery Institute was explicitly united around this goal and they developed strategies of cultural exposure and confrontation to influence public opinion well beyond scientific arguments.[1] These approaches, referred to as "wedge" strategies, continue today with an often acknowledged motivation of reinstating Christian values in education and society. Philip Johnson, widely credited with jump-starting the ID movement, charged that the underlying intent of evolutionary thinking is "to persuade the public to believe that there is no purposeful intelligence that transcends the natural world."[2] Although ID may be presented as if it is only a scientific argument, it is rarely uncoupled from this foundation of attacking evolution as a philosophical worldview (even as many advocates declare otherwise). The significant question for Christians is whether the ID challenge is a proper or effective approach for influencing society with a vibrant Christian message.

Some observers have claimed that intelligent design is simply a repackaged version of young-earth creation science, but this is not an accurate assessment in many respects. Leading ID advocates typically acknowledge that there has been a substantial evolutionary history. Most also explicitly or implicitly accept a very old age for the universe. It follows that intelligent design and creation science leaders are not close allies, even though they are united by their common ground of challenging the prevalence of evolutionary theory.

I appreciate the appeal of ID for many Christians, as it seems to provide a path for demonstrating God within the natural mechanisms studied by science. However, ID does not satisfactorily deliver on this promise. As much as I want, as an evangelical Christian, to proclaim the existence and importance of God, I do not believe the ID movement accomplishes this and it is, instead, detrimental in many ways. I will explain why in this chapter. For Christians who are thinking that all I ever do is dismiss God at every turn, I can tell you I will also begin to discuss how we may successfully search for and reveal God's hand in nature. The ultimate goal is to develop a coherent view of God's sovereignty in his creation that integrates with a vital and engaged understanding of science.

1. Center for Renewal, *Wedge.*
2. Johnson, *Darwin on Trial,* 189.

Intelligent Design as a Scientific Pursuit

First, it should be agreed that the basic intelligent design formulation is a God-centered explanation within the practice of science. ID proponents commonly assert that they are not making a religious argument because the precise nature of the designing intelligence is left unspecified (even alien visitors are discussed as an option). Despite this posturing, the connections to a biblical view of an intervening Creator are transparent and Christian motivations behind the proposals are often clearly stated.[3] To an unbiased observer, an undefined "intelligence" that is large and powerful enough to create complexity or intercede so that there are major gaps within evolutionary progressions (all in an unknown fashion) is a pretty certain conceptualization of a God or Supreme Being. It is also apparent that Christian communities are attracted to ID because it points to God's involvement within scientific processes. Based on my previous discussions of the distinctive nature of science, it is probably apparent that this insertion of God as a scientific conclusion raises prominent red flags for the scientific community.

In many respects, intelligent design is a modern resurrection of the "argument from design" developed by English theologian William Paley during the late 1700s and early 1800s.[4] In Paley's most famous characterization, he reasoned that if someone finds an intricately constructed pocket watch they immediately infer that it was produced by a designing intellect, a "watchmaker." In the same way, the complex, integrated characteristics of living things also imply an organizing designer, a Creator-God. Paley's conceptualization was the prevailing view of the natural world until Darwin and others illuminated natural causative processes that could also generate complexity and adaptive design.

A prominent intelligent design focus has been cellular and molecular intricacies of biology. The most-well-discussed illustrations were first generated from Michael Behe's widely read book, *Darwin's Black Box: The Biochemical Challenge to Evolution.* Behe argued that complex mechanisms, such as the coordinated steps of human blood clotting or the structure and movement of a bacteria's flagellum, could not have evolved by successive modifications of natural selection because earlier versions would not function without each of the existing parts of in place. A critical aspect of "irreducible complexity" logic is the insistence that any change in the arrangement of parts would destroy the operation of a biological feature.

3. For example, the connection is explicit in this book title: Dembski, *Intelligent Design: The Bridge Between Science and Theology.*

4. Paley, *Natural Theology.*

Behe has been countered by detailed demonstrations that less-complex versions of the mechanisms he discusses are present and functional in organisms and that the subparts of his complex systems are sometimes found operating in different contexts within organisms. For example, various blood clotting systems occur across animals with increasing numbers of elements and degrees of complexity.[5] This shows that the primary principle of irreducible complexity is inherently flawed. A succession of evolutionary steps from one system to another, with each intermediate step operating successfully in the organism that possesses it, is an entirely reasonable scientific proposition. In discussions of the bacterial flagellum, biologists have pointed out that many of the protein components of the flagellum structure are found in other functional systems within bacteria.[6] For example, one group of flagellum proteins is used elsewhere in bacteria to inject toxins into neighboring cells. Evolutionary biologists have understood for some time that structures can develop from parts originally functioning in a different context.

Many of the systems Behe and others have used as unexplainable examples of irreducible complexity have been sensibly addressed and given good evolutionary explanations and supportive data for how they could develop by successive modifications. This includes the development of light-gathering structures leading to complex eyes[7] (frequently cited by spontaneous creationists as an unsolvable evolutionary puzzle). Although there has been a substantial body of written materials since the initial intelligent design presentations, both the fundamental ID premise and the response of mainstream biologists have remained the same.[8] There are certainly numerous complex mysteries in the natural world that inspire and challenge all scientists, but the scientific arguments of ID have never convincingly demonstrated the sole action of an intelligent designer or excluded explanations based on natural mechanisms. There have been numerous good resources that have detailed why.[9]

---

5. Miller, *Only a Theory*, 30–69; Shanks and Karsai, "Self-Organization."

6. Musgrove, "Bacterial Flagellum"; Miller, "Flagellum Unspun."

7. Schwab, *Evolution's Witness*.

8. Sample ID titles: Behe, *The Edge*; Dembski, *Design of Life*; Meyer, *Signature*; Meyer, *Darwin's Doubts*.

9. Forrest and Gross, *Trojan Horse*; Miller, *Only a Theory*; Prothero, "Fumbling Bumbling"; Shermer, *Why Darwin Matters*; Young and Edis, *Intelligent Design Fails*; TalkDesign.org.

Is There a Place for ID in Science?

Beyond the difficulties described above, the most significant problem for intelligent design is that it attacks the very foundation of scientific disciplines. In previous chapters, I have discussed the value and limits of natural science and explained that understanding and accepting its parameters is a huge factor in resolving creation-evolution controversies. Scientists do their best to develop exclusively natural explanations, but this does not deny that God exists or suggest that he is uninvolved in his creation. Biologists use and propose evolutionary descriptions because an enormous body of unifying evidence has already confirmed an evolutionary past (acknowledged by the majority of intelligent design leaders). This is true even as we recognize that there are many details we do not understand about evolutionary history.

In contrast, the focus of intelligent design is to suggest we can make a "scientific" conclusion that an unknown designer miraculously intervened in an abrupt fashion to create biological features. Almost all scientists quickly respond that this does not utilize natural processes to explain what is observed and it dangerously confuses and misrepresents the science disciplines. Most ID advocates are eager to insert God as the "designer." But, if the design argument is allowed, we could just as legitimately say an unknown sorcerer somewhere performed an incredible act of magic. In other words, *any* explanation for the designer is acceptable as long as it is *empirically unknowable*. The revealed problem is that design hypotheses do not provide any explanatory power or clarity. There are no natural processes or mechanisms that could ever be critiqued, extended or refined, only the non-dissectible claim that a designer is responsible for complexity in particular biological features. Invoking inexplicable miracles or magic does not attempt to describe or explain biological structures and phenomena the way normal science does. In other words, intelligent design fundamentally redefines what science is and what it is trying to accomplish.

However, intelligent design authors add a philosophical counter that requires careful consideration. They claim that their insertion of a designer (God) to explain biological complexities is no different than mainline scientists inserting evolutionary speculations to explain the same things. They point out that scientists can't always describe satisfactory mechanisms (particularly that natural selection is not adequate) and that some things seem especially "unexplainable" in terms of natural processes. For ID advocates, it follows that conclusions about intricacies in biology are ultimately determined by a person's metaphysical worldview. Authors suggest that an insistence on naturalistic explanations in science is an artificial constraint that inappropriately imposes a materialistic philosophy and excludes God.

In this way, the intelligent design movement complains that we are being forced to conform to a mechanistic approach by atheistic prejudices.

The problem is that offering a design conclusion and offering evolutionary speculations are absolutely not equivalent types of proposals. The difference is that intelligent design is a nebulous, God-centered, special creation conclusion, while evolutionary biology is a continued search for natural explanations. The vast majority of scientists, including both biblical Christians and atheists, reject intelligent design simply because it is bad science. They recognize that ID is nonproductive and essentially stops science in its tracks. Even further, they understand it is deceptive to align ID with a theistic worldview and evolutionary explanations with materialism.

Sadly, those of us who offer the comprehensive view of evolutionary creation are also dismissed as succumbing to materialistic philosophy. We are lumped with all other evolutionary biologists as "Darwinists" and inaccurately represented as if we claim that the mechanism of natural selection is a wholly adequate explanation for everything in nature.[10] Most important, ID authors have falsely portrayed evolutionary creation as if it excludes God from the process of creation.

Redefining Science for a Cultural Benefit

Please let me repeat that I am concerned about atheistic tendencies in scientific communities. All Christians should be united in opposing activities in society that do not allow individuals to affirm their commitment to God and the Bible. I detailed many problems in chapter 7. Even so, concerns that Christians may have about materialistic thinking in science, or in society at large, are not sufficient motivation for distorting the practice of science.

I suggest, along with most Christian colleagues in the scientific community, that intelligent design proposals are more damaging than helpful. ID writers like to point out that too many scientists act like we have a complete handle on evolution. But it is a mistake to think or imply that overzealous statements about natural mechanisms are always an attempt to exclude God or commit us to atheism. Natural science can only make progress if it always pushes on to explain puzzling observations in terms of natural processes. Evolutionary theory does not interject natural selection as an answer to everything. Rather, it says that there is a solid conclusion about a long evolutionary past. In turn, we are actively studying evolution's precise path and working to better understand the known and unknown natural mechanisms that brought it about. In contrast, intelligent design approaches do

10. Dembski, "Is Darwinism Theologically Neutral?"; Richards, "Evolutionism."

not really demonstrate the actions of God or any other designer within science. It is very telling that ID has not won over very many scientists. We must allow for this to be a legitimate rejection of ID without dismissing the response as materialistic bias.

A significant question for Christians is whether the intelligent design wedge has helped counter materialism. I suggest that it has not. What it has done is: (1) increase the divisiveness that already exists between science and faith communities, (2) empowered nonscientists, including politicians, to think incorrectly about science and scientists, (3) interfered with legitimate science education about the diversity of life on earth, and (4) emboldened individuals to believe that they can substitute whatever they would like as scientific conclusions. It is very difficult to conclude that the ID attack on the foundational methods of science has had a positive cultural effect. I ask fellow Christians to consider whether these antagonistic confrontations are actually beneficial.

The obvious alternative for Christians is a more positive engagement of vibrant faith with the undebated consensus of mainstream scientists. Evolutionary creation offers this alternative. As a Christian who enthusiastically pursues Jesus and also loves the search for scientific truths, my heart aches over the ongoing battles between Christian and science communities. From what I have observed, ID is not bringing glory to God, even though that it is what most advocates intend. I suggest that God does not need our help to prove his actions in natural processes when he hasn't already chosen to make himself blatantly obvious with notable evidence of miraculous interventions. There are already solid historical Christian understandings of God's actions in the natural world that do not require or expect abrupt interventions (see below and the next chapter). This means intelligent design logic is unnecessary. I will continue to invite Christians to celebrate the intricate evolutionary history that God orchestrated while simultaneously offering meaningful witness for the gospel of Jesus and a Christian worldview.

Looking For Clarity

There are some revealing problems in the structure of ID proposals that may help us better understand how to move forward. The irreducible complexity argument is framed as if a strong description of "complexity" is positive evidence for a designer and evidence against evolution. But this is not really positive evidence at all. We also know that evolutionary theory provides an explanation (a naturalistic one) for adaptation ("design") and also predicts complexity. So, how much intricacy or complexity is required to conclude that there is evidence for ID?

I'm certain that intelligent design enthusiasts will protest that I am over-simplifying, but the essential feature of the ID argument is that God can be the explanation if biological complexity is *too great* to be explained. The difficulty is that "too great" is not a quantifiable or empirically testable notion. For a simple example, a person in favor of intelligent design judges that a bacterial flagellum is irreducibly complex; it is too complicated to have developed by natural processes. But an evolutionary biologist will look at the exact same scientific information and judge that the flagellum could have reasonably developed through evolutionary mechanisms. Importantly, the two conclusions are not really driven by natural observations or data, but by an individual's understanding of what is acceptable science (sound familiar?).

Some ID authors offer intricate mathematical presentations for "complex specified information" and elaborate arguments for "information" and "significance" in biological structures.[11] However, the depths of these analyses do not alter the essential detail that the purpose is to prove an unattainable conclusion that all possible known and unknown natural explanations have been ruled out. ID arguments have been continually countered by presentations that show it is not logically possible to conclusively demonstrate something was created by an intervening designer rather than by natural processes.[12] The exposed problem for ID is that its a priori stated purpose is to "seek evidence of design in nature."[13]

ID is, in essence, a sophisticated version of an old "God of the gaps" argument that says if we cannot explain something with natural explanations, then God did it. But there is no real value in coming to a specific point of uncertainty or complexity in research and concluding that a miracle happened. This is commonly understood and usually not debated. Instead of using God or miracles as an endpoint, we continue to propose and test naturalistic explanations. I encourage fellow Christians to examine whether our heart's desire to acknowledge God may lead us to too readily embrace this kind of thinking.

Christian scientists could, with very good intent, project God into various holes in scientific knowledge (e.g., inexplicable aspects of developmental biology), but we would be destined to look foolish as new scientific information fills in the supposed unexplainable holes (e.g., a new

11. Dembski, *No Free Lunch*; Meyer, *Signature*.

12. Isaak, "Question of God"; Schneider, "Dissecting Dembski"; Venema, "Seeking a Signature."

13. In the Discovery Institute's definition of intelligent design.

understanding of genes that control embryonic development). The fact that any particular gap in knowledge could even *theoretically* be filled with new scientific information shows that the idea that God accounted for the gap was really no explanation at all. We cannot use God in those cases where things currently seem particularly mysterious or incomprehensibly complex, because science has a history of filling in just those sorts of things. For example, diseases that were once thought to be of mysterious origin are now accounted for by our detailed knowledge of infectious bacteria and viruses. Also, what once were puzzling aspects of organism distribution around the globe are now explained by modern understandings of plate tectonics and continental drift.

As a science educator, anytime I use the phrase, "This is not fully understood," I could also interject, "God may be responsible for this by miraculous intervention." But this imaginary scenario highlights what is really at issue. I would be injecting a religious message without adding any *scientific* clarity about the topic I was discussing. As a Christian I know that all complexities or gaps in our knowledge *could* be due to God's intervention, but as a biologist I recognize there is no scientific value in invoking that explanation. As a *Christian believer*, I embrace the understanding that God has created and sustains all life on earth, but within my profession as a *scientist*, I specifically and purposely pursue scientific explanations for natural phenomena without invoking the direct action of the God I love.

What Are We Arguing for?

Many Christian readers may be frustrated by my analysis. You may be asking how I can say I believe that God is real and then squash this attempt to demonstrate him within scientific discoveries. I must stop and turn this line of thinking around. All Christians believe in "intelligent design" in the sense that God is responsible for bringing about and sustaining the natural world. But why should we expect to demonstrate God in small hidden details of modern science? I suggest that this is inconsistent with what we already know about the God of the Bible and his relationship to his creation. We worship an all-powerful, sovereign God who we accept does not reveal himself to us obviously in physical form (other than in the special case of Jesus). Our common experience tells us that God continually interacts with the world in a way in that is not empirically detectable. Christian theologians, as well as common believers, understand that God has chosen for his people to come to him by a developed faith rather than by surrendering to the presence of indisputable physical evidence. With this experience, why should God be evident only now in irreducible complexities uncovered by advanced techniques of modern investigation? I believe that a critical question arises: why are we looking for a scientifically detectable God at all?

The Bible proclaims that God *is evident* in the natural world. Psalm 19:1 announces, "The heavens declare the glory of God; the skies proclaim the work of his hands." Romans 1:20 asserts, "For since the creation of the world God's invisible qualities—his eternal power and divine nature—have been clearly seen, being understood from what has been made . . . " These truths were apparent, long before scientific inquiry, to those who had the faith to perceive them. The same truths are evident today in the awe, wonder, and beauty of nature and it is through these senses that Christians understand God in nature. We do not need scientific studies to expose God in the natural world and I believe Christians are looking in the wrong place when they anticipate that empirical research will indisputably reveal him. The greatest irony is that as some Christian scientists search for an elusive God found only where science knowledge is lacking, we point people away from a more straightforward perception that he has been apparent all along. I will have much more to say on this subject in the next chapter.

I believe an absolutely crucial conclusion for Christians is that scientific answers are not the key path for revealing or proving God. I plead with my brothers and sisters in Christ to stop looking within the disciplines of science for confirmation that God exists or is important. For many believers, this may need to hit like a striking revelation and be implemented with a complete shift in expectations. Christians should accept that proper science will endlessly seek out natural explanations for features of the natural world. Not only may we allow science to progress in this way, we can excitedly participate in this admirable calling to uncover the details of God's creation. On top of that, Christian can anticipate that scientific discoveries will reveal more and more about the glories of God and inform our sense of worship. Maintaining this clarity of science as solely based on natural descriptions separates out the still vital and deeply important discussion of whether or not there is a God and whether it matters.

The Public Arena

Despite the fact that scientists have provided a strong basis for rejecting intelligent design, its promotion continues to challenge public policy regarding the teaching of evolution. A primary thrust is to claim that ID arguments demonstrate a significant debate among scientists. In the early years, a primary strategy was to suggest that schools should "teach the controversy" so students would be informed about competing scientific views or, at least, about the inadequacies of evolutionary theory. Among a number of public figures, President George W. Bush famously entered the arena by commenting, "Both sides should be properly taught, so people can understand what the debate is about"[14] (while his own chief science advisor, John Marburger, simultaneously noted, "There is no real debate"[15]). At the time of these statements in 2005, new local or state-level challenges to the teaching of evolution were in the works in 20 different states.[16] Most of these efforts have since been dropped or overturned.

The most prominent and influential case came out of Pennsylvania. The Dover Area School Board voted in late 2004 to require biology teachers to indicate that there are problems with the theory of evolution and to include intelligent design as an alternative. A subsequent suit led to a six-week trial in federal district court with expert testimony from ID proponents and specialists from the scientific community. In December 2005, Judge John Jones (a Bush-appointed Republican and Lutheran) ruled in a detailed opinion that intelligent design is "faith masquerading as science." As with the 1980s refutation of creation science, our courts and the scientific community recognize that intelligent design is a flawed scientific challenge to evolutionary theory and it is a philosophical or religious claim rather than a scientific proposition. Judge Jones noted that it was evident to any reasonable observer that the intent of ID is to project God into science classrooms. As both a concerned evangelical Christian and lifelong scientist, I must agree with the judge's conclusions.

Even after this landmark case, the ID movement continues and new local legislative attempts emerge with backing from the Discovery Institute. The original "teach the controversy" strategy has morphed into arguments for "academic freedom" and suggestions that schools should "teach the strengths and weaknesses of evolution." A very telling observation is that these phrases are often voiced without any reference to specific scientific information. The general population hears that evolutionary theory is "inadequate," that there is "scientific evidence against evolution," and that there is good reason for skepticism about the prevailing views of "Darwinism." When given surveys with leading choices, the public objects to education that teaches "*only* evolution" and they support academic freedom to teach "both sides" of the issue. But the poll respondents are agreeing to something

14. Wallis, "Evolution Wars," 28.
15. Alter, "Monkey See," 27.
16. Wallis, 28-29.

that doesn't even make sense to professional scientists. ID advocates have succeeded in creating confusion and doubt about evolutionary biology.

But evolutionary theory does not have "weaknesses" in the way that is being implied and there really isn't any "evidence against evolution" in the natural world (except the problematic rationale of intelligent design). Evolutionary theory is the most confirmed and comprehensive conceptual frameworks in all of biology. We know there is incomplete information about many aspects of the earth's evolutionary past, but that is something that is expected and not a difficulty with the concept. Most importantly, there is no alternative naturalistic proposition to teach students. What is laid bare is that the vocal opposition to evolution is driven by philosophical discomfort with natural explanations that do not include God. Mainstream scientists understand that these are rhetorical battles not scientific ones.

I agree with my scientist colleagues that what our students need most in this realm is a solid education about the process and wonderful explanatory power of evolution. Of course, I also believe that students should be openly encouraged to develop their own metaphysical worldviews. However, I am saddened that so many Christian students are preprogrammed to approach evolutionary biology with reluctance and disbelief. I suggest that this is ultimately not beneficial for our country or the Christian church.

The struggle has played out at some colleges and universities where there are reports of censorship against scientists who promote ID. When scientists have been reprimanded, it has been touted as the scientific establishment restricting the free flow of ideas. Frequently, the harmed individual has been presented in the media or in spontaneous creationist literature as a martyr who is trying only to follow the evidence.[17] It is difficult to assess the specifics in every case, including the real potential for unfavorable treatment toward scientists with Christian beliefs. However, each situation needs to be evaluated in light of the scientific community *defending the purity of scientific endeavors.* Many analyses have demonstrated that claims of martyrdom or persecution are distorted.[18] This situation is really no different than arguments over creation science in previous years. A vast community of both Christian and non-Christian scientists rejects intelligent design because the approach forces religious pronouncements into scientific discussions. I am not aware of a single case of censorship in which this criterion alone is not the legitimate reason for the reported action.

The general population continues to communicate that it does not embrace or, very likely, does not understand the reasons why God should be

17. e.g., the documentary film, *Expelled.*
18. Schloss, "Expelled Controversy."

excluded from the methods of science. A 2005 Pew Research poll indicated that 64% of Americans support teaching "creationism" along with evolution in public schools.[19] The poll did not distinguish what kind of creationism is favored (young-earth creation science? old-earth progressive creation? intelligent design? evolutionary creation?). However, it and other polls like it show that that the majority of individuals do not want God excluded from our public dialogue.

It is my contention that our American society needs desperately to discern how to incorporate outward discussions of God and spirituality in our public lives while also appreciating the distinctive, productive methods of scientific inquiry. I have offered a viable suggestion that science educators could help reduce tension if they were carefully trained to include presentations within introductory courses about the distinctive nature of science and spirituality. Christians, in turn, must utilize effective ways to share their faith and worldview without attacking and disrupting science.

I start my own classes for biology majors with an outward acknowledgement that I am a Christian and that a Creator-God and evolutionary theory can be compatible. I communicate an appreciation and respect for the spiritual decisions each individual makes. I also indicate that a holistic view for each person includes an incorporation of scientific information into his or her own faith-based perspective. I try to create a comfortable environment in which any metaphysical worldview, from atheism to evangelical Christianity, is acknowledged and accepted. I am comfortable pointing out that, theoretically, there can be unexplainable aspects of biology due to the intervention of God (they are called miracles), but also that the methods of science based on natural mechanisms will not be able to detect them. Then I move on and do science throughout the rest of the course. I will typically have later spiritual discussions with some interested students. Even though I enjoy sharing the details of my Christian faith, I do not manipulate my science classroom to do so. It would be unfair and inappropriate for me to misuse my role and authority as a public classroom instructor to invoke God or specifically promote my own spiritual beliefs. Of course, it is just as inappropriate for a science educator to use this setting to promote atheism. What I do know is that students thrive with the scientific wonders of evolution when they know their spiritual perspectives are respected and safe.

19. Pew Research, "Religion."

10

The Pursuit of God and Science

IN THE LAST TWO chapters, I have deflected the attempts of creation science and intelligent design to inject God into scientific discussions of life's diversity. As a Christian devoted to Jesus, my heart understands the motivations behind these proposals because part of me also wants to logically demonstrate God's hand in every aspect of biology. But, I have explained how this will always be a misguided way to proclaim Christian faith and it will invariably be a contorted, unconvincing communication.

Even so, I am absolutely convinced that God is present in the world and in our lives. My primary assertion is that we ultimately know the truth of a living God and a spiritual reality through a sense that is not the same as scientific investigation. I sense the presence of God in nature, in his creation, every day. I am in awe and I am thankful. Christians sing many songs of praise that proclaim God's power, majesty, and beauty in creation and I sing them wholeheartedly. I am certain that he caused the world to come into existence and he sustains it now. He set up the physical parameters of nature. He has guided and joyed in the evolutionary proliferation of organisms over vast amounts of time (billions of years) just as he joys in the uncountable billions of stars that are beyond our comprehension. Just how a God that is before and after time has done and does all this is not wholly knowable. But that is not a copout. Christians believe in a God that we do not physically see and we accept that he has chosen not to reveal himself in distinct form to each of us individually even though he could. In the same way, God does not reveal his actions blatantly in the natural world even though he could. Still, he is obvious to me, by faith, in nature, in Scripture, in my prayer life, in my church family, and more.

The ultimate issue for Christians in creation-evolution discussions is describing a workable understanding of the relationship between God and his creation. How does God interact with the natural world?

A Harmony of Causes

Christians proclaim that God not only created the natural world, but also that he continually maintains it. Nehemiah 9:6 exclaims, "You alone are the Lord. You made the heavens, even the highest heavens, and all their starry host, the earth and all that is on it, the seas and all that are in them. You give life to everything..." Further, Colossians 1:17 declares, "in him all things hold together" and Hebrews 1:3 affirms that Jesus is "sustaining all things by his powerful word." In one sense we can say that God does this through continuous "miracles." This is because what he is precisely doing and how he is accomplishing it is hidden or unfathomable from our limited human perceptions. This view of God's constant involvement is a fundamental aspect of Christian faith.

On the other hand, we know that God's ongoing activity maintains a trustworthy and orderly creation with inherent natural properties. Our common experience tells us that the universe is intelligible and it is governed by consistent mechanisms that can be reliably studied. Because of this, scientists may examine the properties of atomic elements, gravity, the movement of celestial bodies, and the biological mechanisms of life. In one previously used example, we can roll a ball down a ramp, measure its friction, speed and momentum, and predict how far it will travel once it hits level ground. God is not evident in this event even though Christians assert he is responsible for the natural order by which it occurs. Importantly, we absolutely rely on the expectation that God does not intervene when the ball is half way down the ramp to knock it sideways. We may also note that describing details of the rolling ball does not exclude God. A large step in solving creation-evolution confusion is seeing that descriptions of evolutionary mechanisms or the path of evolution is principally no different than these descriptions of the moving ball.

Given the backdrop that God sustains natural processes, what should we expect to see when we go into nature? In what way should we expect to observe God's hand? Scripture indicates that we should see the "glory of God." However, there is no particular reason to expect evidence of obvious miraculous interventions scattered within the orderly creation. Romans

1:20 provides important guidance: "For since the creation of the world God's invisible qualities - his eternal power and divine nature - have been clearly seen, being understood from what has been made . . . " In this verse, Paul communicates that God's "invisible qualities" are obvious and "clearly seen." However, there is no indication that we should see conspicuous marks of God's creative acts. It certainly seems that Paul would have stated this if he believed the natural world contained such distinct signs.

For a satisfying understanding of God's relationship to his creation, we need to throw off mechanistic perspectives that have come to dominate western society and appreciate a robust theological view understood by early Hebrews and upheld through the historical Christian church. In this view we may comfortably embrace that God is the author and cause of events (through his ongoing activities) while simultaneously saying that natural processes also cause the same events. This understanding is sometimes referred to as "dual causation."[1]

First, we know that God is outside of his creation rather than merely being part of it. One way to glimpse this idea is to consider that God existed before he created the natural world out of nothing. John 1:1–3 reads, "In the beginning was the Word, and the Word was with God, and the Word was God. He was with God in the beginning. Through him all things were made; without him nothing was made that has been made." We also speak of God's omnipresence (unlimited by any bound of time). Psalm 90:2 states, "Before the mountains were born or you brought forth the earth and the world, from everlasting to everlasting you are God." What this means is that God is beyond the space and time-defined dimensions of the natural world. This is not an intuitively easy perception and yet it is a key part of Christian theology. Any description of God's interactions with nature must incorporate this understanding.

An imperfect but helpful analogy is to picture the relationship between God and his creation as something like that between a playwright and a play.[2] The author of the play constructs the storyline, determines what will occur on stage, and establishes its conclusions. However, as anyone watches the play, they do not see any direct evidence of the playwright within the action. Importantly, the actors are responsible for and are causing what is occurring as the play progresses, but in a different way, the original

1. Arguments for dual causation are largely associated with Thomas Aquinas although its conceptual roots extend through Plato and Aristotle. For an outstanding detailing of Aquinas' views see Carroll, "Creation, Evolution and Thomas Aquinas."

2. I have borrowed the playwright-to-play analogy with permission from Ric Machuga. His writings contain thorough and helpful explanations of dual causation. See Machuga, *Life, the Universe and Everything*, 247–56.

author is also the cause of the play. In this analogy, the playwright is the primary cause and the actors and stage mechanisms are secondary causes. In the traditional theological view, God is the primary cause, while the living and nonliving components and processes of the natural world operate as secondary causes.

What may surprise many Christians is that dual causation is not simply a philosophical construct; the Bible itself proclaims it! Psalm 139:13 reads, "For you created my inmost being; you knit me together in my mother's womb." Believers respond to this expression of God's control with wonderment and comfort. And yet, modern science can describe how the human form arises from the unfolding of specific genes and progresses through developmental steps without apparent intervention by God. The fact that a baby's formation can be described by observable natural mechanisms does not negate that God caused it to happen. They are both true.

Similarly, Psalm 102:18 states, "Let this be written for a future generation, that a people not yet created may praise the Lord." This verse refers to future humans that will later read the psalmist's words. God will "create" these people, but there is no indication that they will come about by any means other than normal human ancestry. Clearly, God's creative action and natural processes are being pointed at simultaneously and being referred to as the same event.

This leads us to also consider how God's creation of plants and animals might be viewed in a comparable fashion. It is ingrained in many sections of the Christian church that the history of life on earth must involve instantaneous special creation events, but Scripture does not dictate this. In fact, there is plain reference to dual causation within the creation story of Genesis 1. Particularly intriguing is the coupling of verses about the origin of living things. For example, Genesis 1:24 reads, "*Let the land produce* living creatures according to their kinds: livestock, creatures that move along the ground, and wild animals" while the next verse addresses the same events with "*God made* the wild animals according to their kinds, the livestock according to their kinds, and all the creatures that move along the ground according to their kinds" (italics are mine). It is significant that these paired sentences *equate* something in the land with God's creative action. The phrasing, "let the land produce" implies that there is some mechanism or process inherent in the land that can produce animals. In other words, the verses indicate dual involvement of natural causes and God, not instantaneous creation by God alone. You may also notice that evolutionary processes can fit nicely as the implied natural mechanism. I will have much more to say about the Genesis 1 creation account in chapter 17.

Once we grasp a solid view of dual causation, there is no direct reason to impose instantaneous special creation events such as those demanded by young-earth creation and intelligent design. In fact, why should we expect abrupt interventions in natural mechanisms when our known perceptions of the natural world and the Bible itself show us this is not how things work? Our primary difficulty with dual causation is our restrictive tendency to describe activities only in terms of the cause-and-effect relationships humans experience on earth. But our starting point is that God transcends our perceptions of space and time and we must allow for his relationship with his creation to be different in character than the mechanisms we observe in nature.

There is an apparent sense of mystery in dual causation, but that is part of the point. We are talking about the God of the universe who is beyond the limits of our own existence and experiences. The topic is framed incorrectly when we are forced to choose between two opposing options; either that God did it or it happened by natural processes. Explaining an event scientifically in terms of natural causes does not discount that God was the primary cause behind it. Instead, it is all by God's design and his ongoing activity. This means that it is perfectly reasonable for a natural phenomenon to be simultaneously explained by the actions of God and by scientific descriptions.[3] In the context of evolutionary biology, we understand that God utilized his own ongoing natural processes to bring about the diversity of life on earth.

What About Miracles?

Many Christians quickly object that when evolutionary creationists use the dual causation approach we are denying the existence of miracles. This is not the case. All biblical Christians expect that God can penetrate into the natural world in an apparent way to suspend or uniquely affect the natural processes that are the common order. As an evangelical Christian, I positively accept miracles. Since God is the sustaining author of creation, it is obvious that he may at any time unusually influence the consistent patterns he normally maintains. Please notice that miracles are not distinguished by being the only time God interacts with his creation, they are when he chooses to do something out of the ordinary. Miracles are expressed throughout the Old and New Testament and are experienced by believers today.

Some miracles are events distinguishable by the improbable coincidences of timing.[4] Acts 16:25-26 describes an event when Paul and Silas

---

3. Old Testament theologian John Walton provides a "layer cake" analogy to illustrate the concept of dual causation. See Walton, *Lost World of Genesis*, 114–17.

4. Humphreys, *Miracles*.

were in prison singing hymns to God. While they were singing, a violent earthquake shook the building's foundation and the prison doors flew open. It is significant that the people involved did not see God directly. They also would have been able to describe the earthquake as a natural occurrence. But the remarkable nature of the event was evident enough to the participants for it to be recorded as a miracle in the New Testament. Individuals who do not believe that God is present and active can always explain away this type of miracle as a coincidence, while Christians anticipate that God sometimes chooses to operate in this fashion. The ultimate examples of miracles are those that completely defy natural descriptions, particularly the virgin birth and resurrection of Jesus—miracles that specifically communicate God's redeeming love and are the underpinning of Christian faith.

An important question for Christians is to ask when and in what context God chooses to do miraculous events. Biblical miracles are "signs" and "wonders" that are accomplished in situations where God (Jesus) purposes to communicate with a single person or a group of individuals. The message is always a theological one to express the character of God and/or his desired relationship with those who follow him. A miracle is not so much about the event itself, it is about the guiding message associated with it. For example, Jesus' New Testament miracles, such as healing a blind man or changing water to wine, are not simply to give a poor man sight or perpetuate a party. They are explicitly to reveal Jesus and his gospel. Theologian and physicist John Polkinghorne expresses the point in this way:

> Miracles are not to be interpreted as divine acts against the laws of nature (for those laws are themselves expressions of God's will) but as more profound revelations of the character of the divine relationship to creation. To be credible, miracles must convey a deeper understanding than could have been obtained without them.[5]

In contrast, there is no specific indication that God has chosen to use miracles in the normal course of natural history. Evolutionary creationists certainly understand that God could have broken the ordinary sustained processes of nature during the derivation of living forms. However, there are two main reasons to suspect that he has not.[6] First, miraculous events in the distant past without human witness do not fit the pattern of biblical miracles where there is an associated theological message for God's people. Second, there is no indication from the natural world that God worked by

5. Polkinghorne, *Science and Theology*, 93.
6. Louis, "Miracles and Science."

miraculously altering the regularity of natural processes (i.e., no obvious evidence for young-earth creation or intelligent design).

Divine Action

Given the preceding views, how specifically does someone approach life's diversity by simultaneously doing science and believing in an active God? On the science side, we understand that the properties God created in the natural world have led to powerful processes of evolutionary change. We know how genetic recombination, DNA mutations, and natural selection can lead to major changes in the form of living organisms, and the evidences in support of the general concept of evolution suggest that these, and undoubtedly other mechanisms, have been players in forming the diversity of organisms on earth.

At the same time, we understand that God has constantly been in charge of the natural world through his sustaining actions. This includes intimately and undetectably affecting the course of evolution by influencing normal evolutionary mechanisms. These notions are consistent with the God of the Bible who has related with humanity through history, and with whom Christians have an interactive personal relationship. Even so, the exact nature of God's influence or intervention is difficult to discern.

The primary theoretical issue that remains is to ponder the hidden connection between God, as the primary cause, and natural entities and processes, as secondary causes. In the playwright analogy of dual causation, this is like asking how the author's stage play instructions are transferred to the actors of the play. Similarly, we may say that God directs the formation of a human fetus or causes the wind to blow, but we are tempted to contemplate "how" he does it.

An initial response is to proclaim that understanding how a transcendent God is enacting his interactions with his creation is unknowable. The pursuit may be akin to a blind man trying to understand colors. Even so, we can wonder if God guided evolution by causing specific mutations in the DNA of organisms or altering environmental conditions that determine the types of organisms that survive. We can similarly visualize God's hand in continental drift, ice ages, or an asteroid striking the earth. A variety of evolutionary creationists have discussed how they understand God's interactions within the creative processes of evolution.[7] This includes authors who have elegantly explained how God may take action within the indeterminacy of quantum mechanics or within complex systems without suspending

7. Bartholomew, *God, Chance and Purpose*; Polkinghorne, *Science and Providence*.

the discernable operation of natural processes.[8] Modern physics and "chaos theory" demonstrate that knowing all present conditions and parameters is never enough to predict future conditions. In other words, God may act within these systems without humans perceiving or understanding it. In essence, God may operate in undetectable ways within processes that appear entirely random to us.

But, a very crucial point is that none of these suggestions for God's involvement is a scientific proposition. One cannot investigate natural processes to demonstrate where God intervened or affected the natural world. Unless God chose to blatantly leave evidence that he intervened (as ID leader, William Dembski, has suggested, a "made by Yahweh" emblazoned on every cell), we will not be able to distinguish a miraculous event from things brought about by the natural processes he originally created and sustains. Intelligent design writers argue that "complexity" and gaps in scientific explanations are evidence for miracles, but these are indistinguishable from currently unexplained phenomena that resulted from natural causes.

Perceiving God

It is a very important question to ask why God has not left indisputably obvious signs of his involvement in the natural world. Intelligent design authors have lumped evolutionary creationists (theistic evolutionists) with atheist evolutionists for denying that God's creative actions can be scientifically demonstrated. Dembski wrote the following criticism:

> Within theistic evolution, God is a master of stealth who constantly eluded our best efforts to detect him empirically. Yes, the theistic evolutionist believes that the universe is designed. Yet insofar as there is design in the universe, it is design we recognize strictly through the eyes of faith. Accordingly the physical world in itself provides no evidence that life is designed.[9]

Perhaps surprisingly, I am not particularly offended by this critique and I agree with the statement in principle (although I don't view God as a "master of stealth"). What is unstated here is the underlying presumption that God's influence *should* be scientifically detectable in nature. On what basis is this assumption made? God has not made his hand abundantly clear (no "made by Yahweh" signs), even though Christians believe he could, so why

8. Miller, *Darwin's God*, 192–219; Murphy, "Divine Action"; Russell, "Special Providence"; Russell, "God Who Acts."

9. Dembski, "Introduction," 20.

do intelligent design proponents believe they should be able to demonstrate him in subtle complexities of science? Why should we think we are revealing God's hand in the modern intricacies of cellular biology if it wasn't already obvious in the structure of whole organisms? As an evolutionary creationist, I am confident that God has been and continues to be responsible for his natural creation, but I also understand he has chosen to do so in ways that we cannot scientifically delineate.

This discussion is really no different than asking why God does not perform notable miraculous acts in the world today to openly declare his existence and his glory. Those of us who believe by faith do so without expecting any visibly open act from God to the world. We accept that there are good theological reasons (discussed through the history of the church) that he has chosen a relationship with his creation that requires reasoned faith rather than a concession to the obvious. As Jesus declares in John 20:29, "Blessed are those who have not seen and yet have believed."[10] I remain comfortable and confident that God is the Creator and Sustainer of life and that he influenced the mechanistic evolutionary processes he created. It is ironic indeed (!) that the efforts of creation science and intelligent design advocates inadvertently suggest that God is not real or relevant unless he can be shown by scientific methods. This diverts attention from a more spiritual confirmation of God's existence and importance.

To those who perceive it, God's hand is recognized in the majesty and wonder that the natural world inspires and in a sense that there is something more than the physical and biological processes we observe. The Bible pronounces that God's glory and power are evident in his creation and it is significant that this claim is not based at all on gaps in our scientific understandings. All of us sense this power and wonder in nature; the real question is whether we attribute it to God. I found in my studies of the natural world that this sense was intuitively apparent and it was a relief and a joy when I came to a place of accepting that God is behind it all. I discovered after my own conversion that there is a rich history in Christian theology of recognizing "beauty" as the handprint of God on his creation. This is an important path to perceiving the truth of his being. Atheistic scientists and philosophers reason that "awe" and "beauty" are evolutionary products in our brains' neural circuitry, but this is not really a fully satisfactory explanation (see chapter 2). Each of us must choose by a rationale faith to either concede that there is a God behind our existence and the world we live in or decide that there is not.

10. Also see Is 45:15, 1 Cor 13:12.

In this view, there is some irony in that Paley's watchmaker analogy was not altogether inaccurate. Where Paley saw intricacy and design in nature that inspired both wonder and appreciation, I believe he was indeed seeing evidence of a Creator-God. Not evidence for a spontaneous creation event, but signs of the God behind it all. Of course, this view of things can never be confirmed or rejected by scientific study. Perhaps John Calvin said it best in his reiteration of Hebrews 11:3: "Through no other means than faith can it be understood that the worlds were made by the Word of God."[11]

There will always be a tension between spiritual truths and scientific knowledge. While the Bible tells us we are "fearfully and wonderfully made" (Ps 139:14) and that God forms us in our mothers' wombs (Isa 44:24), it is simultaneously true that the development of the human fetus is controlled by an individual's DNA. Similarly, while God certainly intended for us to joy in the splendor of the lilies of the field (Matt 6:27–29), it is also biologically accurate to say that the purpose of flowers to plants is to attract animal pollinators without any reference whatsoever to humans. Any understanding that blends both spiritual and scientific wisdom will accommodate these dualities. Christians should be cautious not to shun scientific knowledge to uphold a simple or mysterious image of God. It is exhilarating and comforting to know that God constructed our body parts in our mothers' wombs and scattered flowers across the landscape knowing that we would enjoy them, but biological knowledge reveals that these perceptions are not a full picture of what is true about God's creation. Conversely, scientists should not reject the spiritual reality that speaks to so many of us by insisting that descriptions of natural mechanisms somehow exclude it. One context in which many of us already work out this duality is when a loved one is sick or injured. We pray because we trust in a wondrous God who is sovereign and has an all-encompassing view of what is going on, but we also trust in the wisdom of doctors who have been equipped by the advances of scientific knowledge.

Here is the bottom line: I know my God is responsible for life on earth. He is sovereign, all-knowing, and all-powerful. I also accept an evolutionary view of the past. How and where God has specifically operated within evolutionary processes to determine the diversity of life on earth is not fully knowable or testable. I am comfortable with this because it is all that we can know. I am also confident that honest scientific investigation will yield increasing understanding about past events in the natural world that will never be in opposition to the God of Scripture.

So, I do science and attempt to explain what we can with natural explanations. In one example, I can observe Winter Wrens in the dense

---

11. Calvin, *Commentary*, Vol. 1, Argument.

underbrush and House Wrens in the canopy of trees and analyze how past events within a common ancestral wren population might have led to the evolutionary development of these two species. As I use scientific thinking, I am not acting as if God does not exist; rather, I am confirming my belief in a natural order that God created. At the same time, I proceed with amazement at the splendor and beauty in the world and ponder how God has and does operate. I am thrilled by the antics of the two wrens as they move through the vegetation and my heart is joyful as the sunlight bounces in multicolor off the foliage. I do my work "heartily as for the Lord rather than for men" because it is the Lord Jesus whom I serve (Col 3:23–24). I am at peace because my God provides meaning and purpose to all that is around me and all I do. My analysis of the evolutionary history of the wrens is in the realm of science, but every bit of my existence, including the science I practice and the worldview I use to make sense of it, is guided by and given context by my Christian faith.

As a Christian who accepts the power and sovereignty of God and as a biologist who accepts the scientific integrity of evolutionary theory, I am guided to evolutionary creation as a comprehensive view of past events on earth.

PART IV

The Value of Biological Evolution

PEOPLE OFTEN ASK, "WHY are biologists so persuaded that evolution has occurred to produce the diversity of life?" The straightforward answer is that nature presents abundant and varied *evidence* that is difficult to account for if it is not the result of evolution. But there is something far more organic and exciting than this academic statement. That is, once the idea of evolutionary change is considered, we find that it has enormous power to explain much of what we see in the biological realm. This *explanatory power* pervades all levels of biology, extending from the origin of cell organelles to complex interactions within ecosystems. With evolution in mind, phenomena in one area of biology after another become understandable like they never were before. Biologists are able to repeatedly exclaim, "Aha! I get it! Now, that makes sense." It is because of this sweeping power to make sense of the natural world that evolution is regarded as one of the few unifying principles in the biology discipline. As a Christian biologist, these evolutionary explanations lead to comparable exclamations of, "So that's how God did it!" In the following chapters, I will survey a sample of evidence for biological evolution and provide illustrations of what I mean by its explanatory power. These pages also contain many lessons about evolutionary processes, about the world we live in, and how we might rightly view it.

Any discussion of evolutionary evidence is accompanied with a dilemma. I am aware, of course, that there are many protests from spontaneous creationists. In an internet Google search for "evolution" or a similar entry, a good number of the top responses will be young-earth creation or intelligent design websites. These sites present aggressive arguments to negate or diminish evolutionary theory and they promote the notion that there is a lively debate over its validity. There are also large amounts of printed material from spontaneous creationist organizations. Major effort is devoted to rebutting the evidence presented in high school and college textbooks and to discrediting the latest discoveries announced in the scientific literature. The arguments are so wide-ranging that entire books have been offered to challenge the evolutionary paradigm in science classrooms.[1] Significant parts of the general population accept these assessments and communicate their displeasure that evolutionary biology is so strongly defended in academic communities.

On the other hand, I have already noted that scientists have considered and responded to spontaneous creationist challenges. The National Center for Science Education is the most consistent defender of evolutionary science. Its efforts are easily accessible online.[2] This organization stays abreast of each anti-evolution publication, presentation, and political movement and offers articles and public responses that expose the difficulties contained within these challenges. The NCSE is not at all focused on religious compatibilities. Even so, it is generally sensitive to issues that evolution poses for Christians and the organization allows that a biblical God and evolution can be harmonized. Most important, the NCSE explains how the evidence for evolution is deep and rich, and how mainstream scientists routinely use evolutionary theory to foster productive research without even a hint of debate about its reliability.

With this backdrop, I will present selected examples of evolutionary evidence with the qualification that I'll be presenting only a fraction of what's available. I will also address the key implications of some spontaneous creationist critiques. Even so, I am mindful to not let this book be derailed by attempting to address every criticism. First, these critiques have been effectively handled elsewhere (e.g., by the NCSE). Second, although critics can readily claim that I am avoiding their arguments, scientists have learned that to be drawn too deeply into these kinds of interchanges inadvertently continues an air of credibility for arguments that we really feel have been clearly dismissed. This encourages the false sense there is a genuine controversy when there is not.

I especially have no desire to argue with fellow Christians about the soundness of evolutionary theory. As I stated at the very beginning,

1. Davis and Kenyon, *Of Pandas*; Meyer et al., *Explore*.
2. See NCSE.com. Critiques of spontaneous creationist textbooks include, Matzke, "Of Pandas"; NCSE, "Explore."

nonproductive exchanges have gone on for far too long and they deflect our focus from far more important constructive conversations. Instead, I am eager to continue with both Christians and non-Christians who sense that scientists have genuine reasons to conclude that the earth is very old and life is united through ancestral connections. My purpose in this section is to help readers more fully appreciate why these are robust and meaningful conclusions. It is from this position that we can discuss what it all means, especially for Christians with a strong biblical worldview.

My illustrations will provide enough diversity and detail to reveal a reasonable picture of the evolutionary landscape seen by biologists. Since I offer this information as an evangelical Christian biologist, I will address spiritual and cultural implications as well. I especially hope to communicate a scientifically accurate view of evolution. I have seen that evolutionary concepts are often dismissed because of superficial or confused assessments of what biologists are really saying. If you are not biologically inclined, I encourage you to stick with this unfolding of information because it is so essential for establishing a true understanding of what evolutionary biology is all about. It really is wonderfully fascinating and tells an intricate and astonishing story about the history of life on earth.

Ultimately, I have two goals. The first is to communicate a sensible understanding of the biological foundation behind evolutionary theory. The second is to continue to express how someone may accept that the biological world is both the product of evolutionary processes and the intended creation of a sovereign God.

11

Living Architecture

Consider for a moment what the animal world would look like if each species was fully unique and no two kinds shared any similarities. This is, of course, impossible to imagine. Instead, what we typically see in nature is groupings of animals or plants that share comparable forms and common structures. This is the basis for familiar categories such as snails (species with a slimy gliding foot and a spiral shell), insects (animals with an outer skeleton, three pairs of jointed legs and often two pairs of wings), and vertebrates (forms with a backbone surrounding a spinal cord), to describe just a few. There are, in turn, subgroups within these larger categories with their own definitive features, such as true flies or beetles among the insects and birds or mammals among the vertebrates.

These basic observations have been the grounds of dispute between evolutionary biologists and spontaneous creationists since the days of Darwin. Evolutionists, including evolutionary creationists, say that the shared architectural plan within each group or subgroup was present in one original ancestral species and the plan was modified in different ways during the formation of descendant species. Many spontaneous creationists, particularly young-earth creationists, counter that God repeatedly used common body plans, or archetypes, while designing and creating members within the particular groups. How can we distinguish between these two options?

Comparing Forms

The evolution section of biology textbooks almost always contains an illustration of vertebrate forelimbs like that in Figure 11.1. This shows the front appendages of different non-fish vertebrates (i.e., amphibians, reptiles, birds, and mammals). A close look reveals that the same basic bony elements are present in every animal. These include a single bone in the upper limb (humerus), two side-by-side bones in the "forearm" (ulna and radius), multiple small bones in the "wrist" (carpals), and then a series of long thin bones through the hand and digits (metacarpals and phalanges). These same bones are present whether it is the sprinting leg of a cat, the flipper of a whale, the grasping arm of a human, the front leg of a lizard, or the wing of a bat or bird. Although the bones are in the same relative positions, their individual shapes and sizes differ greatly because of the structural criteria required for limbs of such varied functions. The simple visual imagery of this illustration is one of the most straightforward and compelling arguments for the reality of an evolutionary past.

If evolutionary thinking is excluded, it is difficult or impossible to give a scientific explanation for why the forelimb bones should be the same in animals of such different form and styles of locomotion. Certainly, there is no particular reason to *expect* that such functionally varied forelimbs would contain the same bones inside. However, when ancestral connections are considered, the forelimb patterns suggest that an early ancestor (a primitive amphibian in this case) possessed the original bony framework and it was variously modified in different evolutionary lines to fit the requirements for different types of locomotion. Biologists refer to these vertebrate forelimbs as *homologous structures*; that is, similar structures in different species that are thought to have been derived from a common ancestral condition.

The alternative offered by some spontaneous creationists is that the Creator simply used the same basic bones and relative position of bones in all amphibians, reptiles, birds, and mammals when they were created. The difficulty with this approach is that it is merely a statement and not an explanatory framework for what is observed. The internal structure of specially created forelimbs could theoretically take a huge variety of possible forms and no biologists would *a priori* predict that the forelimb bones would be the same in animals as disparate as a frog, a cat, and a bird. The spontaneous creationist "explanation" amounts to a statement that God just wanted to do it this way.

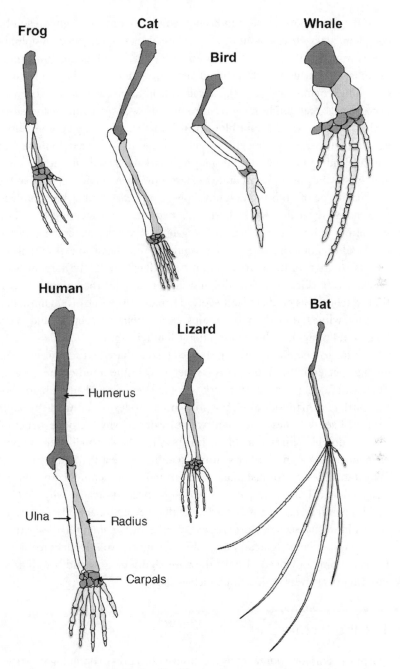

Figure 11.1 — A sample of vertebrate forelimbs. The same bones are present in each of the examples although there are differences in the shape and sizes of the bones.

The important message in homologous structures is strengthened by complementary observations of what is known as "convergence." A dolphin is a swimming animal that is also a warm-blooded, air-breathing mammal. There is good evidence from shared anatomical and physiological features (e.g., lungs, mammary glands), as well as from molecular data and transitional fossils, that dolphins evolved from land-dwelling mammal ancestors. A shark in contrast is a cold-blooded fish that shares many commonalties with other fishes, including gills. What is intriguing is that a dolphin flipper and the leading fin on a shark display a remarkable match in *external* form and the way in which they operate as these animals swim. It appears that the external shape of both the dolphin flipper and shark fin has been molded to a shared form that works best for a hydrodynamic steering limb in a swimming animal (they have "converged"). Given the matching external form and function of these limbs, it would be reasonable to expect that the *internal* support of these limbs might be similar as well. They are, instead, very different (Fig. 11.2). The dolphin flipper contains the bones of terrestrial vertebrates as predicted if it evolved from the forelimb of a mammalian ancestor, while the shark limb contains a series of base structures and many long, thin fin rays like those found in most other fishes.

Prior to Darwin, all biologists recognized shared structural patterns among species as common plans (archetypes) in the mind of the Creator. There was no real motivation to ask why God would build amphibian, reptile, bird, and mammal forelimbs for varied modes of life with the same restricted set of bones. But ought we not wonder why the bone structure within a dolphin's fin is more like a lizard's or monkey's than like the support elements inside a fin of other swimming animals like a shark or other fish? Why didn't the Creator just build each extremity with a unique, specifically designed set of bones? And, if the limbs were specially created, why did God build them in a way that is consistent with an evolutionary interpretation? An evolutionary scenario provides a *scientific explanation* for these structures that spontaneous creationism cannot. I am an evolutionary creationist because I am convinced of both the sovereignty of God and the value of meaningful answers to questions such as these.

The Big Picture

The most common spontaneous creationist objection to homologous structures is that they are simply based on assumed evolutionary relationships or circular reasoning. For example, some contend that an octopus eye is not considered homologous with a human eye *only* because it is already

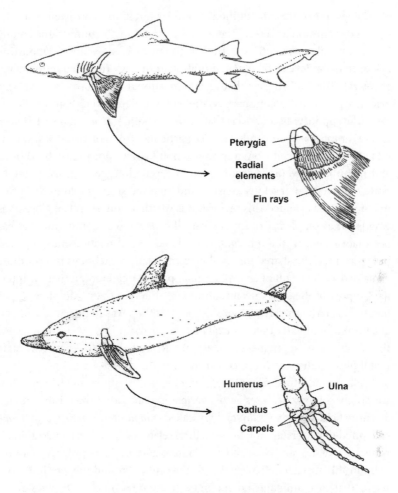

Figure 11.2 — Convergence in the forelimbs of sharks and dolphins. The external form of the shark pectoral fin and the dolphin flipper are quite similar because they are both used for the same function; to steer the body through the water while propulsion is provided from the rear. The internal support structure of the limbs, however, is very different and suggests that the limbs had different evolutionary origins.

assumed that octopuses and humans are not related.[1] This is a misrepresentation and it is misleading. Instead, biologists conclude that there is a homology *only* through a total picture constructed from a collection of information.[2] Land vertebrates share a huge number of anatomical and

---

1. Wells, *Icons*, 59-66.
2. Gishlick, "Icons"; Max, "Homology"; Matzke, "Icon."

physiological features, in addition to similar forelimbs, consistent with the notion of common ancestry. There are also good fossil connections among fishes, amphibians, reptiles, birds, and mammals, as well as supporting evidence from developmental and molecular biology (detailed in following chapters). These all suggest that vertebrate animals are closely related. This evidence points to evolutionary relationships as depicted in Figure 11.3.

The supported scenario is that (1) fishes were the original vertebrates, that (2) amphibians descended from a particular group of *air-breathing* fishes, (3) reptiles evolved from a specific amphibian group, and then (4) birds and mammals developed from different reptile lineages. I must pause to point out what a marvelous, elegant, and involved story is told in the information that has yielded this picture! It is worth asking skeptical Christians whether the complexity of the story itself is likely to be something that has been fabricated. In the full context of the ancestral relationships depicted in Figure 11.3, the shared forelimb pattern in land vertebrates is beautifully explained as the result of an evolutionary past. We can reason that the internal support of the shark pectoral fin looks different than a dolphin flipper because a shark is a cartilaginous fish and its limb was not derived from an ancestral amphibian. There is *explanatory power* in this evolutionary view that is absent in spontaneous creationist formulations that can only state that it pleased the Creator to construct the animals this way.

Numerous other examples of homologous structures are evident when we closely compare features across vertebrates and other organisms. In fact, an entire field of study within biology, called *comparative vertebrate anatomy*, is devoted to understanding the structural relationships of vertebrate animals from an evolutionary perspective.[3] Courses and graduate programs are offered in this specialty at colleges and universities around the world. Recognizing that each animal is not a unique creation has intellectually rewarded biologists and we have learned that a full understanding of animal form comes from seeing each species in its evolutionary context. This is a good place for evangelical Christians to hover for a moment. How are the many thousands of professionals and students in comparative vertebrate anatomy supposed to take the objections of special creationists seriously?

As an additional example of vertebrate homology, investigations of reptile scales, bird feathers, and mammal hair reveal that these structures are evolutionarily connected. All three develop in a similar fashion from depressions in the outer layer of the skin and all are composed of the same primary structural protein called keratin. These uniting features make sense

3. Kardong, *Vertebrates*; Kent, *Comparative Anatomy*; Liem et al., *Functional Anatomy*.

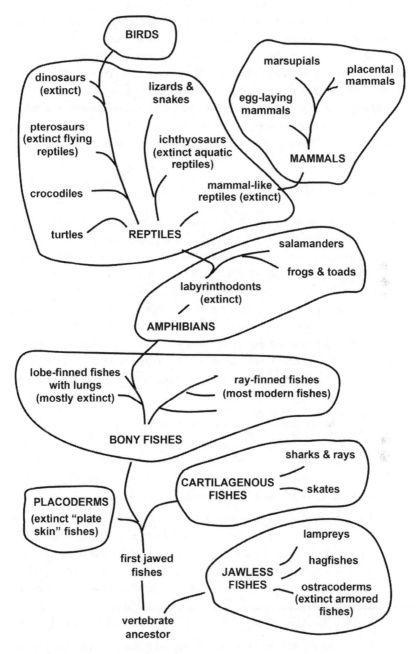

Figure 11.3 — An evolutionary tree for vertebrate animals. These relationships are based on various lines of evidence, including comparative anatomy, developmental biology, molecular comparisons, and transitional fossils between groups.

in an evolutionary view because both bird feathers and mammalian hair are understood to have evolved from keratin-based scales on reptile ancestors (see Fig. 11.3). Again, this conclusion is not based on the study of scales, feathers, and hair alone, but from a total picture of corresponding evidence from developmental and molecular biology and the fossil record. Biologists can match and compare an array of corresponding traits across the vertebrate groups, including comparable bones, muscles, heart features, blood vessel patterns, digestive organs, and reproductive structures.

Structures without Function

Particularly compelling evidence for an evolutionary view is the presence of homologous structures that are rudimentary, and typically functionless, in the organisms that possess them. One aspect of nature that fascinates all observers is that organisms have remarkable features that adapt them for their particular way of life. Just a few examples are the long thin bill of a hummingbird for extracting nectar from deep inside tubular flowers, the camouflaged form of a walking stick insect, and the delicate tufts on a dandelion seed that catch the wind to disperse the seed. A more surprising thing is that nature also reveals complex structures in organisms that *serve no purpose at all*. These make no sense in organisms that were specially created by God for specific modes of life.

Consider the following:

- Most moles are blind for life in accordance with their underground existence. However, they also have the basic form of an eye, including a bony eye socket, eyelids, a structural "eyeball," and even a lens, but a functioning retina for sight is not formed.

- The hooves of elks are composed of two functional digits that contact the ground (as in deer), but there are also two small dangling digits further up the side of the foot (one on either side) that are never used in the animal's locomotion (Fig. 11.4). Hoofed animal fossils with four toes touching the ground are well-documented and there are other living hoofed species, such as pigs, that show an intermediate condition of four functional toes but with the side toes considerably reduced in size.

- Boa constrictors have two small knobs with bony internal support off the posterior region of their bodies in the same region where elongated lizards have two functional hind limbs.

- Flowers of the genus *Penstamen* form five stamens; four of them have elongated stalks and terminal anthers that yield pollen, but the fifth is always an isolated elongated stalk without an anther (it is sterile). Flowers in similar plant groups possess five typical stamens, all of which produce pollen.

- Flightless beetles grow perfectly formed wings that remain permanently locked away under their fused wing covers.

- Additional examples include rudimentary pelvic bones embedded in the bodies of whales where hind limbs are expected in mammals, stubby wings in many flightless birds, and partially formed eye structures in blind cave fish.

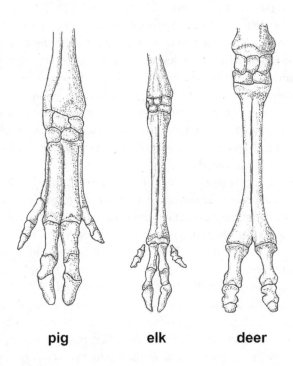

pig   **elk**   **deer**

Figure 11.4 — The front feet of three hoofed animals. The pig foot (left) has four toes, with the side toes smaller than the central ones. The elk (center) has two functional toes, but also two tiny incomplete side toes. The latter are explained as vestigial structures left over in the evolutionary progression from a four-toed ancestor. The deer (right) shows the condition of only two functional toes.

What is common to these and numerous other examples is that they are functionless structures in organisms that clearly resemble (i.e., are homologous with) useful structures in similar organisms. How can we account for these observations? Evolutionists explain the features as remnants of fully developed structures that were present in an ancestral past. They refer to them as *vestigial structures* (vestige = remains, a visible sign left by something no longer existent—*Webster's Dictionary*).

Vestigial structures are a particularly strong argument for evolution because they demonstrate a *lack* of adaptation. Both spontaneous creationist and evolutionists expect to see adaptation in nature. A spontaneous creationist says a Creator designed specific features for an organism's particular existence. An evolutionist states that evolutionary mechanisms selected organisms that possessed the best possible features from a collection of alternative individuals. Both should result in adaptation or "design" in individuals within populations. However, evolution is expected to produce adaptations only from the best available options. Vestigial structures are imperfections that are explainable as remnants of an imperfect evolutionary process. They are entirely unexpected in specially created organisms designed by God. Vestigial structures are particularly perplexing in the context of intelligent design logic. They are not irreducible complex structures that couldn't have evolved by natural processes; they are functionless features no designer should ever bother to construct!

Spontaneous creationists have tried to dismiss the challenge of vestigial structures in a couple of ways. The most common approach is to suggest that all such structures actually have some unknown function in the organisms that possess them or to say that we cannot know the intent of a designer in creating such things. This would mean God had some reason to create them. To most observers this is an unconvincing attempt to salvage a special creation viewpoint rather than a meaningful defense. It is a weak argument to claim that all the very numerous examples of vestigial structures *may* have a purpose while being unable to demonstrate in most cases what those purposes are or what they might be. What possible function could rudimentary eye parts serve for a mole that never opens its eyes or sealed wings serve for a flightless beetle? All this aside, it is not always necessary for a vestigial structure to be entirely functionless for it to support evolution.[4] The dangling side toes in an elk might be necessary as ligament attachment sites in that animal, but they are still obviously reduced and unused *toes* that don't contact the ground during locomotion. In the same way, whatever function might be discovered for the tiny, stubby wings in flightless birds,

4. Senter, "Vestigial Structures"; Theobold, "29+ Evidences," Part 2.1.

it is still evident that they are a reduced, residual form of a *wing* with the anatomy that serves the purpose of flight in other birds.

A different spontaneous creationist response to vestigial structures is to acknowledge that they are true remnants of ancestral organs but to claim that the organs lost their function after the Fall of Adam and Eve. The primary ironic difficulty with this suggestion is that the structural changes and lost functions are supposed to be due to enormous amounts of *evolutionary change* over the *exceedingly brief time* period since that biblical event. There is little evidence that evolution can proceed at a pace like this. (See my further discussion of "baraminology" in chapter 13).

The most straightforward interpretation of the natural evidence is that homologous structures, including vestigial features, are the result of descent with modification. Homologous structures are important support for an evolutionary scenario, but they are only one part of a larger picture. Please read on.

Unfolding Forms

The study of animal development, also known as *embryology*, is the examination of animal growth from a fertilized egg to mature adult form. It may seem odd at first glance to be discussing this topic in the context of evolution. It turns out that the popular practice of embryological studies prior to Darwin revealed some odd surprises and ultimately an evolutionary message. As most animals develop, they progress through embryonic and/or larval stages that look noticeably different than the final adult product. A key revelation is that these stages frequently contain temporary and often functionally unnecessary features that appear to reflect primitive ancestral conditions. Consider the following three examples from a long list of possible illustrations.

(1) Terrestrial salamanders are animals that live in moist locations, such as under logs and in leaf litter. Significantly, they never enter open water to swim. Their development occurs entirely within enclosed eggs deposited somewhere in their moist environment. However, the embryonic stages of these animals possess frilly external gills and a laterally flattened tail fin. These features are just like what is seen in the larval stages of other salamander species in which larvae swim freely in open water. Terrestrial salamander embryos complete a metamorphosis to adult form without ever living outside of their eggs. The gills are reabsorbed, the tail is reshaped and these salamanders emerge with functional lungs and a cylindrical tail adapted for their existence on land.

(2) Embryos of baleen whales, such as a gray whale or humpback whale, form a number of structures within the mother's womb that have no connection with requirements in the functioning adult. Many develop external hind limb buds (complete with various leg bones, nerves and blood vessels), rudimentary external ear lobes, sparse body hair, and obvious teeth. Each feature is visible for a time but then disappears as the embryo grows larger. In these whales, teeth are reabsorbed and replaced by the horny baleen for which this whale group is named. All early fossil forms and most living whales and dolphins have jaws with apparent teeth. Baleen whale embryos also start off with nostrils toward the tip of the snout, but, as development progresses, changes in the shape and size of skull bones cause the nostrils to migrate to their final place at the top of the head to form the blowhole. Various fossil species of cetaceans are known with nostrils at locations ranging all along the front of the head from near the tip of the snout to the position of the familiar whale blowhole.

(3) Finally, most adult snails and slugs have a very strange (to us) body arrangement with the anal opening of the digestive system located at the right side of the *neck* region. Their digestive tract essentially takes a U-turn as it is followed from mouth through the animal and back to the anus on the neck. This anatomy is established during development by a curving growth of the body form called "torsion" (Fig. 11.5). However, there are some slugs that have a different arrangement, most notably colorful sea slugs called nudibranchs with which many people are familiar. Nudibranchs have their anal opening at the back of the body where we typically expect to see it in animals. The odd thing is that nudibranchs get to their adult form by a remarkable sequence of developmental stages. They first go through torsion, just like all other members of the snail/slug group, but as growth continues, the turn in the digestive tract is *undone* and the anus is returned to the rear of the animal where it had started. This latter event is appropriately referred to as "detorsion."

The three developmental examples described here share the fact that the animals form nonessential (and often nonfunctional) features during their development that are recognizable as the useful anatomy of similar living and past species. Evolutionists interpret these patterns as remnants or artifacts left over from modified developmental pathways that were present in ancestors. In parallel with vestigial structures in the previous section, these are *vestigial developmental patterns*. Spontaneous creationist formulations cannot really explain why specially created organisms should have nonessential developmental features that resemble functional structures in similar species.

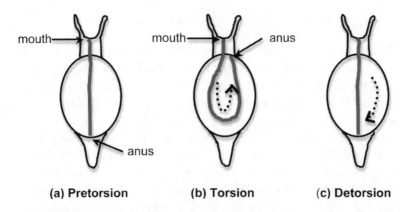

(a) Pretorsion　　　　**(b) Torsion**　　　　**(c) Detorsion**

Figure 11.5 — Body torsion in snails and slugs. (a) Early larval stages start with the mouth in the front and anus in the rear. (b) During further development, the upper body twists 180 degrees to place the anus at the front of the body. Most adult snails and slugs have this anatomical arrangement. (c) In nudibranchs (sea slugs), the larva first goes through torsion, but then the twist is undone to place the anus at its original position in the rear.

But evolutionists understand that terrestrial salamanders evolved from an ancestral salamander species that had aquatic larvae that swam in open water where they used gills for respiration and a tail fin for propulsion. Gills and a tail fin are never really needed inside the eggs of terrestrial species, but they are retained without harm as part of the previously established development stages of all salamanders.

In the same way, baleen whale embryos have hind limb remnants, tiny ear lobes, and sparse body hair because they are artifacts in transforming developmental instructions from terrestrial mammal ancestors that had fully produced versions of these features. The embryos also show nostril migration and teeth because baleen whales evolved from cetacean ancestors with nostrils at the tip of the snout and functional teeth in their jaws. In other words, evolutionary changes in whale developmental instructions have left an embryonic route through some ancestral conditions rather than a path straight to the adult form.

Finally, nudibranchs with an anus in the rear appear to have evolved from a slug ancestor with torsion. Detorsion can be explained as an evolutionary modification of a once-existing slug ancestor that had torsion in its larval development but later experienced evolutionary selective pressures for the torsion to be undone.

The Real Place Where Evolution Happens

The developmental features I have mentioned are consistent with evolutionary scenarios in which new animal forms come from alterations of preexisting organisms. In an evolutionary view, we expect that new features or ways of life will be built on the foundation of previous features and ways of life. Evolution is often characterized too simply as some bizarre series of transformations in an existing *adult* form (such as a lizard sprouting wings to become a bird). This is a distorted misconception that leads to abundant confusion, and it is no surprise that people reject such a notion. But biologists wound never suggest such a thing! Instead, any evolutionary modification that might occur in an adult form is necessarily due to a change in the *developmental instructions* that give rise to that animal. In other words, modern evolutionary theory includes the understanding that changes are really due to alterations in the genetic packages that control the way animals grow from fertilized egg to adult. This is a very important point! It is *only* a change in the genetics of development that can lead to fundamental differences among adult organisms.

With this framework we understand that baleen whales could only evolve by alteration of the existing developmental genes of toothed whales. The existence of non-functional teeth in baleen whale embryos suggests this is exactly what happened (further supported by the fact that baleen whales show up later in the fossil record than toothed whales). Evolutionary theory predicts that there should be obvious connectedness in development pathways of similar species because the genetics of newer species necessarily results from modifications in the genetic packages of older species. Evolutionists would not expect baleen whales to have entirely novel growth patterns different from all other whales.

A simple analogy might be helpful. We can imagine that the development of an organism is something like following assembly instructions for a machine or toy and assume that there are 100 steps in the process. An evolutionary change should be like changing a particular instruction at step 14 or step 82. These could make a small or major difference in the end product of the assembly. The fact that so many of the other steps are still the same should mean that we would still see noticeable similarities between the two assembly processes and final products. Real animal development involves far more steps, and many steps that unfold concurrently with complex interactions, but this analogy can still communicate the essential idea. Evolutionists expect that changes in developmental programs are the ultimate source of much of the novelty in organisms upon which natural selection has operated. Alterations in many of the steps of an established assembly process would fatally disrupt the end product, but others could yield subtle to profound alternatives with adaptive advantages.

An evolutionary view provides a strong explanatory framework for the developmental features I have mentioned. It is difficult to account for them from any spontaneous creationist perspective. This is a difficulty both for those who believe that God specifically and individually created each species (or kind), as well as for individuals who believe God designs complex animal features. Why should animals grow embryo structures that appear to be non-essential or are clearly never used and then later destroyed? Why should these structures directly correspond to functioning structures in similar living or fossil species? Why should animals go through indirect, energy-consuming developmental patterns that seem to reflect ancestral conditions (such as nostril migration in baleen whales or torsion and detorsion in nudibranchs) rather than growing directly to the adult form? We are once again faced with information that is consistent with an evolutionary view of life, but is hard to align with special creation. As a Christian, I am compelled to understand that God used evolutionary transformations to accomplish his purposes.

I must stop to marvel at these things. Unlocking the details of how embryos form strikes me as the unraveling of a wonderful mystery novel. In this case, the story is one of intrigue about the course of evolutionary change. I am amazed at the suggested path of events. The Bible declares that God has made all things. I proclaim it from my knowledge of God's sovereignty but I also joy in the added depth provided by the intricacies of an evolutionary view.

A Subtle and Intricate Example

I am sometimes asked if there is much evidence for evolution outside of animals. The answer is an unqualified "yes." I believe animals are most frequently discussed simply because they are more relatable and understandable for most people. I complete this chapter with a final illustration from a part of biology with which most people have little familiarity. This tale conveys some significant messages.

Biologists have long proposed that the first plant species to live on land evolved from some type of aquatic alga. From an evolutionary perspective, the movement of plantlike organisms from water into terrestrial settings is a very significant event in the history of life, an event that parallels the transition in animals from air-breathing fishes to the first amphibians. Algae are familiar to most people as seaweeds along rocky seashores and as bright green masses at the margins of freshwater lakes and streams. Biologists suggest that particular species from among the algae evolved adaptations necessary for survival on land (such as strong weight-bearing cells and a waxy water-retaining covering) that allowed them to benefit from the abundant carbon dioxide and sunlight available in the aerial atmosphere. Scientists visualize that this "conquering" of bare terrain ultimately led to the plant-filled terrestrial habitats with which we are familiar today.

Following this scenario, plant scientists went looking for clues to decipher the real players in the water-to-land event. First, they had good reasons to expect that plant ancestors would be found among the green algae that populate freshwater lakes and streams. Both land plants and green algae share an array of structural and biochemical features that are unlike other photosynthetic organisms, including specific kinds of shared pigment molecules.[5] The pigments typically make these forms "grass green" rather than various colors exhibited by other photosynthetic organisms (e.g., reddish, golden, brown, or blue-green).

But the question then becomes, where among the green algae is the true ancestral lineage that led to plants? With an evolutionary model, botanists predicted that there should be one limited subgroup within the green algae with distinctive traits linking them most directly to terrestrial plants. Some of the most revealing information comes from what may seem unlikely sources: (1) the anatomy of swimming sperm, and (2) the final steps of cell division. I recognize that these are obscure things to nonscientific individuals but please bear with me because their obscurity and intricacy is ultimately part of their relevance. Figure 11.6 will help a lot.

It may be a fun surprise to learn that *sperm cells* are present in many plants and most green algae. They function in reproduction, just as they do in animals, by swimming through available water from a male structure on one individual to an egg on another. The simplest land plants, including mosses, horsetails, and ferns, produce mobile sperm for fertilization during rainy, wet seasons. Plants that people are most familiar with (such as pine trees, rose bushes, and grasses) bypass the need for swimming sperm through the transfer of pollen grains.

The vast majority of green algae have sperm cells that are spherical in shape and possess two or more *forward projecting* flagella that move the cell rapidly through water (Fig. 11.6b). In addition, most green algae divide one cell into two by forming a new partitioning cell wall that grows *from the outside cell walls inward* until it is completed at the center (Fig. 11.6a). In many species, the process is directed by a collection of long, thin protein tubes at the center of the cell that are oriented *perpendicularly* across the dividing cell.

5. Graham et al., "Phylogenetic Connections"; Graham, *Origin*.

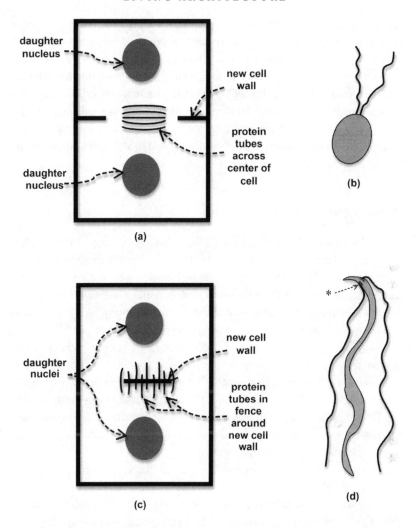

Figure 11.6 — Cell division and sperm cells in green algae and plants.
(a) Cell wall formation in most green algae. (b) The spherical sperm cell of most green algae. (c) Cell wall formation in charophycean green algae and plants. (d) The linear, irregular-shaped sperm cell of charophycean green algae and primitive plants such as mosses, liverworts and ferns. The asterisk (*) indicates the multilayered structure. See the text for an explanation of features.

However, sperm and cell division characteristics are markedly different in one subgroup of green algae, called the "charophyceae." The features in these algae match up directly with what is present in the simplest plants. First, the sperm of both charophycean green algae and plants have a very odd asymmetrical, linear form and they bear two or more *laterally* projecting flagella (Fig. 11.6d). There is also a very distinctive arrangement of layered material inside the sperm cell where the flagella are anchored, simply known as the "multilayered structure." The multilayered structure is never present in the sperm of other green algae. Second, during cell division, a new partitioning cell wall begins in the *center of the cell* and expands outward until it forms a complete boundary between two resulting daughter cells (Fig. 11.6c). This process is directed by a unique arrangement of protein tubes oriented *parallel* to the length of the dividing cell and encircling the developing cell wall like a series of fence posts.

What's the Point?

These observations indicate a strong affinity between the aquatic charophycean green algae and terrestrial plants. This matches the evolutionary prediction that a limited group of green algae should share indicative traits with land plants as a sign of their ancestral connection. Cell division and sperm cell features provide evidence consistent with the notion that a particular lineage within the charophyceae gave rise to the first plants.

Further support comes from analyses of DNA gene sequences. These demonstrate that DNA patterns in simple plant species are more similar to charophycean algae than to any other form of alga.[6] This is exactly what is expected if the genetic instructions of the charophyceans was modified into the DNA instructions of the earliest plants. Evolutionary biologists comfortably conclude that the first land plants evolved from a green alga species within the charophyceae subgroup.

As a scientist, I cannot help but be excited that such an intriguing story can be revealed in the intricacies of biology! Okay, you may not be a biologist and you may be thinking I am a little odd. But the fact that I can tell this story based on biological minutia in organisms from an inconspicuous corner of the natural world is precisely the point. I am aware as a Christian that we should be cautious about questioning God's motives. That said, shouldn't we wonder why God would bother to produce by special creation or intelligent design such subtle features that are so consistent with an evolutionary interpretation? And why might God do so in species that are of particularly little importance or significance in the everyday life of humans? A spontaneous creationist view cannot provide a scientific explanation for these features but evolutionary biology can.

With our unaided eyes, charophycean green algae are not different from other types of green alga, as they display the same types of growth

6. Karol et al., "Closest Living Relatives."

forms and occupy the same diversity of freshwater habitats. It is only when microscopic cellular features are examined that the distinctive nature of the group and the evolutionary message is revealed. This account helps illustrate how deeply into the study of biology the evidence for evolution pervades. Evolutionary theory is not some product of a philosophical bias; it is a repeatedly confirmed framework by which we can make sense of the world of living things.

Finally, this narrative is instructive in another very important way. An evolutionary paradigm prompted plant biologists to look for a particular ancestral lineage to land plants from within the green algae. It drove them to predict that certain algae species would bear distinctive features uniting them with plants. The evolutionary view not only allows scientists to explain what we see, but, as seen in this case, it successfully directs them to ask good questions and make relevant predictions. Part of the strength and worth of a theory is that it does just this. Many spontaneous creationists continue to attack evolutionary theory as if its primary significance or role for biologists is to establish a particular worldview. Its real value is that it routinely guides biologists in doing successful science. Evolutionary theory has proved to be robust and valuable not as a stagnant set of ideas but as a dynamic framework for understanding and further studying the biological world on earth. This is why biologists rightfully defend it as a major guiding paradigm in science today.

12

The Fossil Narrative, Gaps, and Hard Work

IF SOMEONE ASKS ABOUT evidence for biological evolution, the first thing
most people probably think of is fossils. This makes sense because fossils are
the most direct information we have about past events in the history of life
on earth. Scientists define a "fossil" as any physical evidence of past life. This
usually means mineralized remains in sedimentary rock, although an insect
embedded in amber or dinosaur footprints in a hardened streambed also
will do. In our own time frame, we see layers of sediments collecting at river
mouths, in mudflats, and on lake bottoms and ocean floors. In general, the
process is very slow. Yet, exposed sedimentary rock formations in areas of
the Grand Canyon are over a mile thick and in one old lakebed in Africa the
underground sediments exceed a 10-mile depth! These observations alone
suggest to geologists that the earth is very old.

There are a number of reasons scientists expect fossils to be relatively
rare. First, a fossil is formed and discovered only after a series of unlikely
events. The body or a part of a dead organism must be buried quickly, before
being consumed by scavengers or decomposers, and then remain undis-
turbed long enough for the fossilization process to occur. The fossil must
then survive tremendous pressures and geologic processes such as uplifting,
volcanic action, and erosion. Finally the specimen must become exposed
at the earth's surface without destruction and be found by a person capable
of recognizing its significance. These are simple issues of probability that
mean we should not find lots of fossils. Scientists are even less likely to find

fossil remains from older sediments that have been subject to the most wear, from organisms that lived only in localized populations restricted to a small geographic area or from creatures that were relatively rare in the past. In addition, there should not be fossils for organisms that lived in drier, upland environments or other habitats where sedimentation and fossilization do not occur. Even given these constraints, paleontologists have described an estimated 300,000 different types of fossil organisms.

Fossils Are Sequenced

One thing that became apparent to early students of fossils from the mid-1700s into the 1800s was that these remains occur in repeated patterns through sediments. Certain species are typically found together in one sedimentary layer while other species predictably occur in layers above or below. By the middle of the 1800s, geologists had pieced together discoveries from numerous fossil deposits to construct an overall sequence of geological periods (Fig. 12.1). The complexity of fossil evidence undeniably implied that a diversity of organisms had lived on earth in *succession* over a distant past.

At first, geologists tried to make calculations about the age of different sediment layers based on observed rates of sedimentation. They came up with time periods ranging from thousands to hundreds of thousands of years before the present. Today, scientists utilize radiometric dating, a method that relies on the constant radioactive decay of specific chemical elements. By measuring within volcanic rock the ratios of radioactive isotopes and the elements into which they decay, it is possible to determine how much time has passed after the rock cooled from its initial molten state. Volcanic materials deposited among layers of sediments provide time markers for estimating the date of surrounding fossils. This allowed geologists to more accurately attach specific time values to the geological periods developed earlier from fossil sequences. The calculated dates now range into many millions and even billions of years into the past. Although radiometric dating is repeatedly contested by young-earth creationists (which should be no surprise), the principles behind the technique have been thoroughly examined and the practice is well-established and unchallenged in the mainstream scientific community, as well as among old-earth creationists.[1]

---

1. Falk, *Coming to Peace*, 62–73; Giberson and Collins, *Science and Faith*, 63–67; Weins, "Radiometric Dating."

Period (millions of years ago)	Animal Events	Plant Events
Quaternary (2–present)	Extinctions of large mammals	
Tertiary (2–65)	Spread of mammals, birds, and insects	Radiation of flowering plants
Cretaceous (65–140)	Mammals diversify Continued radiation of dinosaurs	Flowering plants begin spread while other seed plants decline
Jurassic (140–205)	Dinosaurs dominate First birds appear	Non-flowering seed plants dominate, especially cycads
Triassic (205–250)	First mammals appear Early dinosaur types dominate	Non-flowering seed plants dominate
Permian (250–285)	Reptiles radiate Amphibians decline	Non-flowering seed plants radiate: cycads conifers, ginkgoes Decline in seedless plants
Carboniferous (285–355)	First reptiles appear Amphibians diverse	Extensive forests of seedless horsetails, ferns and lycophytes
Devonian (355–415)	First amphibians late in period Spread of jawed bony and cartilage fishes	Diversification of advanced seedless plants
Silurian (415–440)	First jawed fishes Diversity of jawless fishes	Simple seedless plants appear on land
Ordovician (440–500)	Diverse invertebrates No vertebrates	Diverse aquatic algae No land plants
Cambrian (500–545)	Representatives of many animal types	Diverse aquatic algae No land plants

Figure 12.1 — The Geological Time Scale

Several conclusions can be made from an overall study of the fossil record. First, life is very old. The earliest evidence of life dates to around 3-1/2 billion years ago. This is an amount of time that is not easily comprehensible for most of us!

Second, most forms that have existed in the past are now extinct. We see many interesting, and even bizarre, fossil organisms, but generally only the more recent sediments contain forms that resemble species alive today. This suggests a dynamic past with many changes over time. Just one example of how fossil organisms can be so different from the familiar creatures of today comes from a 2003 South American discovery of a giant *rodent* the size of a buffalo! This animal, estimated to have lived 6 to 8 million years ago, was unearthed from a region where there are also fossils of giant crocodiles, large marsupial cats, and enormous carnivorous birds.[2]

Third, there are general trends through the sediments. These trends are sequences in organism types that suggest a progression of forms through time, and it is not unreasonable to assume they imply a *process* of progression. Large-scale landmarks for biologists include evidence of the first single-cell organisms with eukaryote cell structure (complex larger cells other than bacteria) at around 2 billion years ago and the first multicellular life forms in sediments ranging from 675 to 750 million years ago.

There are also trends in fossil types within specific groups such as the vertebrates and land plants (see Fig. 12.1). In the backboned vertebrates, we see a sequential emergence and diversification of different animal types through sediment layers from oldest to youngest. This sequence can be followed visually in the evolutionary tree in Figure 11.3 of the last chapter. The pattern starts with (1) strange jawless fishes, and moves through (2) many types of jawed fishes to (3) amphibians to (4) various types of reptiles, especially dinosaurs, and finally, to (5) mammals and birds.

In plants, a comparable succession over time goes from (1) small extinct seedless plants to (2) more complex seedless plants (such as ferns, horsetails, and an important group called lycophytes) to (3) the first kinds of plants that produced seeds (e.g., cycads, ginkgoes, and conifers) to (4) flowering plants. In general, the plant fossils indicate that each group's diversity and distribution waned with the appearance and spread of the next group. From the study of living species, it is evident that this represents a sequence of increasingly efficient adaptations for survival, particularly for successful existence on dry land.

By combining fossil discoveries from sediments of similar age, it is possible to form pictures of what things might have looked like at different

2. Sanchez-Villagra et al., "Largest Extinct Rodent."

times in the past. For example, the Carboniferous period (355 to 280 million years ago) was characterized by moist forests of tree-sized ferns, horsetails, and other non-seed plants The primary land vertebrates at this time were various kinds of amphibians (some up to 5 feet or more in length and most of them unlike the salamanders and frogs we know today). In contrast, the Jurassic period (205 to 140 million years ago) had dry landscapes dominated by large seed-producing trees and shrubs that did not make flowers, especially cycads. The main land vertebrates were many different types of dinosaurs. Scenes of various geological periods like these are often depicted in "dioramas" at natural history museums. It is probably worth pointing out that illustrations appear rather frequently in books and newsprint that misrepresent the organisms that could possibly have occurred together based on what we know from the fossil record. A dinosaur-filled landscape that also includes fields of wildflowers, horses, and humans may sound fun (or scary), but there is no scientific evidence that such a thing has ever existed.

Intermediate Traits

Finally, despite consistent spontaneous creationist claims to the contrary, there are many transitional fossil series demonstrating a progression through animal or plant species with intermediate characteristics. This is just what is expected if species are connected through an evolutionary past. This is not to say that "gaps" between fossil types don't exist. In fact, paleontologists acknowledge that in many situations transitional fossils are still uncommon and they have wondered themselves why more have not been discovered (see below). Nonetheless, there are so many cases of fossil sequences that reveal intermediate features that I cannot begin to explain them all.

There are excellent vertebrate paleontology textbooks and online sites that provide a remarkable catalog of fossil series *between* and *within* all the major vertebrate groups.[3] For example, Kathleen Hunt offers a particularly extensive online documentation for mammal species.[4] This includes transitional series (1) from an ancestral group of reptiles (synapsids) to early mammals, (2) across mammalian orders (e.g., from hoofed mammals to whales), and (3) within smaller taxonomic groups (e.g., extensive fossil series for horses and for elephants). The sheer volume of listed fossil species demonstrates how much more extensive our knowledge of transitional forms is than what is indicated by spontaneous creationist literature or

3. Benton, *Vertebrate Paleontology*; Carroll, *Paleontology and Evolution*; Prothero, *Evolution*.

4. Hunt, "Transitional Vertebrate Fossils."

understood by the general public. Some of the most well-recorded and celebrated transitional series are across major macroevolutionary boundaries
between vertebrate classes, transitions that are especially rejected by most
spontaneous creationists.

One example of a major transition is a collection of fossils that address
the transformation from fish in water to amphibians in terrestrial environments. This alteration in animals has held particular fascination in our cultural perceptions of evolution. I am sure that many readers can say that they
have seen some kind of cartoon characterization of a fish hauling itself out
of water up on to land.

Amphibian forms first appear as fossils during the late Devonian geological period. The most abundant freshwater fishes at this time (based on
the total number and distribution of fossils) were not the common fish species with which we are familiar today. Instead, these fish gained oxygen in
two ways: one, by acquiring it from water using typical fish gills and, two, by
obtaining it from the atmosphere by gulping air into *lungs*. The presence of
both gills and lungs within these animals is itself an intermediate condition
that bridges the generally perceived gap between fish and amphibians.

Members of one subgroup called "lobe-finned fishes" display a number of features that distinguish them as amphibian ancestors. Most notable,
they had unusually strong bone supports and muscles within their fleshy
fins. In addition, they had nostrils that connected through to the mouth as
in all land vertebrates (but not typical in fish) and a distinctive "labyrinth"
tooth structure, with complex infoldings on the tooth surface, found only
in these fishes and many amphibians. The fact that we can describe these
distinctive shared features between one particular fish type and the earliest
land dwellers also significantly links fish and amphibians.

A considerable series of fossil discoveries since the late 1980s has filled
in gaps in our existing knowledge about evolutionary changes from lobe-
finned fish to amphibian.[5] In particular, they have demonstrated that the
standard limb bone arrangement of land vertebrates developed inside of
fleshy *fish fins* on transitional animals that also had both fish gills and lungs
(Fig. 12.2a). In other words, the limb structure of amphibians appeared first
while these animals were still living in water, probably scurrying around in
shallow pools.[6]

A key understanding, in an evolutionary sense, is that the weight-
bearing limb apparatus evolved because it was an effective adaptation for an

5. Ahlberg and Clack, "Firm Step"; Coates and Clack, "Fish-like Gills"; Daeschler et
al., "Devonian Tetrapod"; Shubin et al., "Pectoral Fin"; Shubin et al., "Pelvic Fin."
6. Clack, "Getting A Leg Up."

aquatic animal in shallow water. However, this strong limb would, as a secondary byproduct, open the door for the animal to utilize terrestrial habitat around the water margins. Since there are advantages to spending most of the time out of water (e.g., avoidance of predators, better accessibility to food items), individuals with progressively stronger appendages would survive well. In this way, evolutionary selection would have led to stronger limb support and, eventually, to the loss of unused and unnecessary gills.

The process described here contrasts with a mis-framed evolutionary notion that a fish with gills and weak, inadequate fins went directly onto land and then evolved lungs and weight-bearing limbs because they were *needed* in the terrestrial environment. With this scenario, a skeptic can easily dismiss evolution by declaring that the intermediate transitional form could not possibly survive. This is true, of course, but the dismissal is based on something an evolutionary biologist would never suggest. Instead, we see that many types of fishes first possessed a transitional condition of having both gills and lungs and, then, strong limbs evolved by the progressive modification of functioning fish fins.

Another especially well-recorded vertebrate transition is from dinosaurs that ran on their hind legs to modern birds with the marvelous ability to fly. Numerous existing fossils show key evolutionary changes in skeletal features, including (1) the fusion of many bones along the vertebral column, (2) notable reductions (such as elimination of teeth and the bony tail), (3) lengthening of arm and hand bones with a specialized swiveling bone in the wrist, and (4) the progressive enlargement of bone surfaces for flight muscle attachment (especially, the progressive development of a broad, keeled sternum). It is important to note that these summary observations are based on a very large collection of various fossil species, not speculation based on a few isolated, fragmentary findings.

In addition, some of the most exciting fossil discoveries in recent years have been full-blown feathers and other feather-like structures on over a dozen different dinosaur species.[7] These dinosaurs share numerous anatomical features with birds, but clearly did not have the skeletal apparatus for flying. In other words, even though feathers were present, the animals were *flightless* (Fig. 12.2b).

7. Chiappe, *Glorified Dinosaurs*; Norell and Ellison, *Unearthing Dragons*; Zhang et al., "Fossilized Melanosomes."

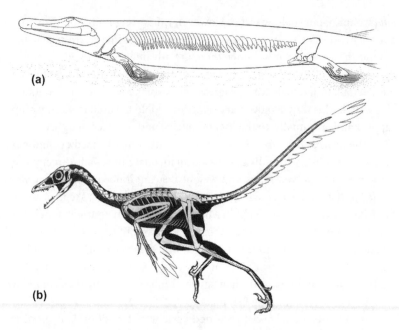

(a)

(b)

Figure 12.2 — Transitional fossils. (a) A simplified reconstruction of *Tiktaalik*, an example fossil species with features between fish and amphibians, including both fish gills and lungs, and strong, supportive limb bones within fish fins. (b) A reconstruction of *Sinornithosaurus*, an example fossil species with characteristics of both bipedal, running dinosaurs and birds. Feathers are present on parts of the body but the skeleton does not show adaptations that would allow flight. The teeth, long bony tail, and unfused backbone are reptilian features, while feathers and the elongated arms and hands are birdlike. [(a) From Shubin, N.H., E.B. Daeschler and F.A. Jenkins, Jr. "Pelvic Girdle and Fin of *Tiktaalik roseae*." Proceedings of the National Academy of Science 111 (2014) 893–99. Used with permission from the National Academy of Science. (b) *Sinornithosaurus* skeletal drawing by Marco Auditore and others. Source: http://fossil.wikia.com/wiki/File:Sinornithosaurus_skeletal1.jpg].

The overall collection of transitional fossils, especially dinosaur species with feathers, completely blurs the generally presumed distinction between reptiles and birds. They indicate either that feathers are not unique to birds or we must expand our notion of "bird" to include some very large, feathered dinosaurs that were adapted only for running! This blurring of taxonomic categories established by humans is just what evolutionary theory predicts. The fact that we can trace a progression of functionally logical changes in skeletal features through fossil species is consistent with an evolutionary view and it is difficult to explain away from spontaneous creationist perspectives.

The discovery of ground-dwelling dinosaurs with feathers indicates that feathers functioned first in some capacity other than for flight, such as for insulation or for display purposes in courtship or aggression. This means that the use of feathers as flight surfaces was a secondary adaptation. We can offer reasonable speculations for how this might have happened.[8] For example, bipedal dinosaurs with feathers on their free forelimbs could gain some lift when they whipped their specialized arms and hands forward to snatch prey. If this minor lift was advantageous in capturing elusive insect prey and/or avoiding predators, individuals with the best feathers would survive most frequently to produce offspring. Further modification to the coverage, size, and shape of feathers could occur over future generations as flight provided additional survival advantages. As with the fish-to-amphibian story above, a feature that evolved first for one particular advantage was then modified to a second highly consequential benefit.

Gaps in the Fossil Record

It is important to address the fact that there are still many discontinuities or "gaps" between discovered fossil forms. Realistic scientists have acknowledged fossil discontinuities and grappled with why there are not even more transitional fossil species that possess intermediate characteristics. Naturalistic solutions have been offered.

One explanation is based on the simple rarity of fossils themselves. Some scientists argue that the chance discovery of a relatively small number of random fossils from all the creatures that have existed is not likely to produce good transitional sequences. This approach is based on the assumption that fossil gaps are essentially the result of incomplete sampling.

A second explanation comes from a more direct (and perhaps more honest) acceptance that the nature of the fossil record represents something real about the biology of organisms. Not only are there existing fossil discontinuities, but additional observations tell us that a good number of fossil species actually remain rather static in form over long stretches of geological time. These realities were addressed head-on with an explanation introduced in the 1970s known as "punctuated equilibrium."

The key idea is that new beneficial adaptations are most likely to arise and spread to new generations within relatively small isolated peripheral

8. Padian and Chiappe, "Origin of Birds"; Prum and Brush, "Which Came First."

populations, rather than within large widespread populations. It makes mathematic and logistical sense that new genetic novelties can infiltrate and become part of the makeup of a small group of interbreeding individuals much more readily and quickly than they might in a population of numerous individuals spread out over large geographic areas. This means that it is very possible that substantial evolutionary modifications have most often occurred within isolated smaller groups, and then individuals with a new form or adaptation have multiplied out geographically to establish new successful wide-ranging species.

Under the punctuated equilibrium scenario, significant evolutionary alterations happen in small collections of individuals while large general populations possess characteristics that stay more stable and static over time. If this is an accurate picture of evolutionary processes, it follows that most discovered fossils will be samples from large static populations while the likelihood of capturing transitional samples from peripheral groups will be relatively small. In other words, our existing fossil record may represent mostly snapshots of widespread successful species and we miss many of the transitional forms that occurred in the past. The beauty of the punctuated equilibrium proposition is that it makes complete sense biologically and it explains our observations of the natural world. This is what all good scientific hypotheses do.

To spontaneous creationists, these suggestions are concocted explanations to cover up apparent gaps between forms they believe were individually created by God. I completely understand this sentiment because it is undeniable that scientists persist in looking to natural explanations for fossil discontinuities. But this is actually an excellent situation for bringing further clarity about the practice of science. A search for an evolutionary-based account is exactly what scientists *should* be doing given our prior support for an evolutionary view (the various lines of evidence that I'm covering in these chapters). We have a solid, useful theory for explaining the diversity of life and we should continue to utilize it and expect it to provide a framework for understanding what is currently unknown unless something of *greater explanatory power* can be proposed to displace it.

Special creation does not provide that alternative because it has continually come up short with regard to natural observations. To insert God as an explanation for fossil gaps is inconsistent with what we know from other biological information. As discussed earlier, it is a precarious position to make theistic conclusions on the assumption that holes in our current knowledge cannot be filled. This is especially true since new intermediate fossils with features predicted by evolutionary theory are discovered on a regular basis.

While I acknowledge the considerable "gappiness" of the fossil record, I believe the transitional series that do exist are critical in resolving this issue. If there are instances of good fossils with intermediate features that support an evolutionary sequence, then it is certainly more difficult for someone to argue that other gaps indicate spontaneous creation by God. In the pages that follow, I will describe one fossil series to more deeply demonstrate how fossil discoveries can reveal a fascinating picture of a dynamic past. This sequence may be familiar to some, because parts of it have been reported in biology textbooks and publically challenged by spontaneous creationists, but it remains a particularly revealing illustration with meaningful insights and lessons.

Returning to the Sea

A collection of remarkable fossils illuminates the fabulous but seemingly unlikely transition from land-dwelling mammals to aquatic cetaceans. Prior to the 1990s, there was a relatively large gap in the fossil record for this transformation, but discoveries since that time have drawn particular attention to it. Biologists have believed for quite awhile that whales and dolphins were derived from land-based ancestors because of their mammalian traits, such as breathing air with lungs and nursing their young with mammary glands. This is, of course, quite unlike the fish with which these animals share their aquatic environment. There have been other clues to an ancestral land existence, such as those features mentioned in the last chapter: vestigial pelvic bones in adult whales, plus earlobes, hind limb buds, and hair that temporarily appear and then degenerate during embryonic development.

Historically, the proposed land ancestor to whales was some species from an early group of now-extinct mammals called mesonychids (Fig. 12.3a). These animals sound very bizarre compared to mammal species with which we are familiar today. They were superficially wolf-like carnivores, but they had unusually large heads and their limbs ended in long toes with *hooves*. Mesonychid fossils occur in sediments ranging from 37 to 60 million years ago (MYA), but there is a particular abundance during the earlier part of this timeframe, before the appearance of any whale-like fossils. A strong link between mesonychids and early whales comes from similarities in the shape and construction of their skull and a close similarity in the form of their teeth.

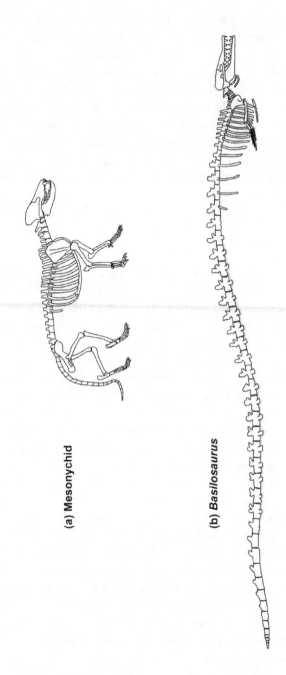

(a) Mesonychid

(b) Basilosaurus

Figure 12.3 — Important fossils in scientist's early understanding of whale evolution: (a) a *Mesonychid* and (b) *Basiolosaurus*.

Until relatively recently, the best-known early fossils of whales were of *Basilosaurus*, which was first discovered and described during the 1800s (Fig. 12.3b). This was a very large, serpent-like animal with a long snout, teeth like other early whales, and a single nostril opening, or blowhole, located part way up the front of its face. *Basilosaurus* was unlike modern whales in a number of respects, but it was clearly a cetacean. The different fossils of this animal have been dated to 35–41 MYA.

The prevailing speculation through much of the twentieth century was that mesonychids were the ancestral source for whales on one branch of an evolutionary tree and that a separate lineage led to all modern hoofed animals (ungulates). Figure 12.4 visually depicts these relationships. With the working notion that a mesonychid ancestor may have evolved into something like *Basilosaurus*, paleontologists predicted the discovery of transitional fossil forms between the two. They also expected that intermediate forms would be found in sediments dated to the time frame between mesonychids and whales (somewhere in the range of 40–55 MYA). In a more general sense, an evolutionary scenario suggested that there should be fossils with traits intermediate to land mammals and aquatic whales; in particular there should be whale-like animals with legs.

This proposition had been a favorite target for many spontaneous creationists up to and through the 1980s. They understandably asked where the transitional fossils were. If such large animals existed in the past in aquatic habitats where fossilization should have been relatively likely, why were there no known intermediates? One representative quotation, which stresses a repeated notion that intermediate forms cannot be functional, illustrates the use of the whale evolution problem in spontaneous creationist literature:

> Darwinists rarely mention the whale because it presents them with one of their most insoluble problems. They believe that somehow a whale must have evolved from an ordinary land-dwelling animal, which took to the sea and lost its legs . . . A land mammal that was in process of becoming a whale would fall between two stools—it would not be fitted for life on land or at sea, and would have no hope of survival.[9]

Since 1989, there has been a burst of new fossil species with intermediate characteristics that have helped fill the gap between land mammals and cetaceans. Each of these has been consistent with an evolutionary transition from hoofed mammals to a semi-aquatic existence and then to fully aquatic whales. I do not hide the fact that I am describing this particular case of transitional fossils because it so beautifully supports evolutionary theory

9. Hayward, *Creation and Evolution*.

and simultaneously illustrates the weakness of claims that intermediate fossils are missing and will not be found. There are many spontaneous creationist challenges, of course, which question evolutionary interpretations made from these remains. However, whatever validity there may be in any of the critiques, the meaningful observation that emerges, as you will see, is that numerous fossils have been discovered possessing intermediate characteristics predicted by paleontologists. I will emphasize four main highlights from among a larger collection of important discoveries.

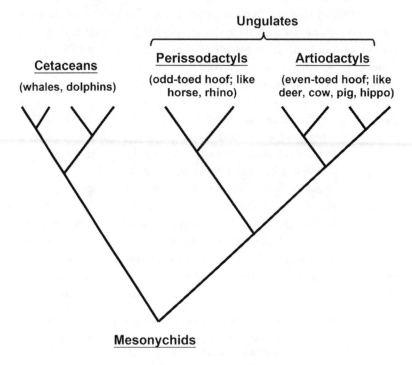

Figure 12.4 — The prevailing view of whale ancestry through most of the 1900s. A group of fossil species, the mesonychids, was seen to be the ancestral stock for *cetaceans* (whales and dolphins) on one lineage and for hoofed mammals (*ungulates*) on a second, separate lineage.

First, in 1989 came the report of a new skeleton of *Basilosaurus*, the well-known ancient whale, but this time with previously undiscovered hind limb remains.[10] The limb had a full anatomy of thigh, leg, ankle, and toe bones (Fig. 12.5), but it was very small relative to the size of this huge whale, and did not appear to have been joined to the backbone (as hind

10. Gingerich et al, "Hind Limbs."

limbs are in other mammals). This means it could not have been much use for locomotion in either swimming or movement on land. This piece of the puzzle shows an early whale with a reduced hind limb (either vestigial or of unknown function). It is consistent with the proposal that whales evolved from land mammals. The finding is also in harmony with our knowledge of modern whales, which have extremely reduced pelvic bone remnants embedded in their bodies, but *entirely* lack external hind limbs. Additional discoveries of ancient whales like *Basilosaurus* with small external hind limbs have confirmed that this condition existed in other fully aquatic early whales (e.g., *Dorudon*).[11]

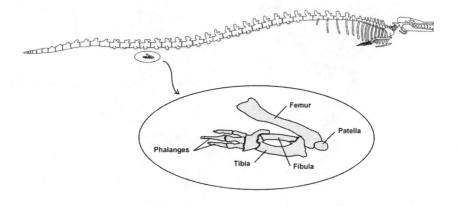

Figure 12.5 — The hind limb of *Basiolosaurus*.

Next, a remarkable fossil was discovered in Pakistan in 1992.[12] *Ambulocetus*, dated to 47–48 MYA, was the size of a large sea lion and had four well-defined legs (Fig. 12.6a). As with many uncovered fossils, knowledge of this animal is based on an assortment of remnant bones rather than a complete skeleton. Even so, contrary to spontaneous creationist objections, quite a lot can be deduced from the features of the bones by comparison to whales and other mammals. *Ambulocetus* showed clear cetacean characteristics in its skull, including a long snout with teeth like early whales (plus traits less obvious to those untrained in whale anatomy, such as a reduced zygomatic arch and enlarged, poorly attached tympanic bullae). This animal's limbs were relatively short, but they ended in large feet, especially the very large hind feet. The shape of the lower back (lumbar) vertebra indicates that the rear spinal column was flexible in an up-and-down vertical direction,

11. Gingerich and Uhen, "*Ancalecetus*."
12. Thewissen et al., "Fossil Evidence."

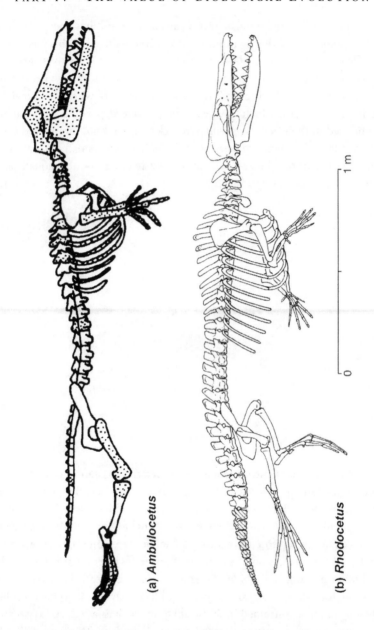

Figure 12.6 — (a) *Ambulocetus* **and (b)** *Rhodocetus*. [(a) From Thewissen, J.G.M., et al. "Fossil Evidence for the Origin of Aquatic Locomotion in Archaeocete Whales." *Science* 263 (1994) 210–12. (b) From Gingerich, Philip D., et al., "Origin of Whales from Early Artiodactyls: Hands and Feet of Eocene Protocetidae from Pakistan." *Science* 293 (2001) 2239–42. Both (a) and (b) reprinted with permission from the American Association for the Advancement of Science].

while prominent processes on the upper back (thoracic) vertebrae indicate strong muscles for accomplishing this vertical action. The one unearthed tail vertebra suggests that there was no fluke or significant tail muscula-ture. Particularly fascinating were the long toes on the hind limbs, each of which ended in a *hoof!*

The anatomy of *Ambulocetus* implies that it was a swimmer that moved with an undulating up-and-down action of its spine and propulsion from its large hind feet, much like sea otters and river otters of today. *Ambulocetus* had limbs that were probably strong enough to support its weight on land, but the lack of prominent attachment sites for walking muscles means that it likely did not move efficiently out of water. It has been suggested that *Ambulocetus* came onto land for resting and mating like modern sea lions. *Ambulocetus* is truly a transitional fossil with features of both whales and *hoofed* land mammals and its anatomy implies that it possibly lived a semi-aquatic existence intermediate to those two mammal types.

An additional fossil discovery came in 1993, also in Pakistan, of *Rodhocetus*.[13] This form is from sediments dated at around 46–47 MYA. It showed many whale-like features, like all the fossils in this discussion, especially in attributes of its well-preserved skull. *Rodhocetus* also displayed characteristics suggesting that it swam with up-and-down motions of its tail region. First, it had very large processes projecting from the lower back vertebrae, suggesting the attachment of strong muscles for moving the tail. Second, its sacral vertebrae were unfused rather than linked together into a single unit as in most mammals. This is an anatomy shared with modern whales that allows flexible movement in the hip region of the spine. Finally, several discovered tail vertebrae were particularly large and robust. These features, in combination, may mean that the animal had some version of a tail fluke. The pelvis and thigh bones of *Rodhocetus* were smaller than in *Ambulocetus*, suggesting they were not well-adapted for bearing weight. It has been suggested that *Rodhocetus* may have been able to locomote on land, but its hind limbs were probably not adequate for sustained locomo-tion there. Later discoveries of additional *Rodhocetus* remains allowed the composite restoration shown in Figure 12.6b.

Finally, a 2001 report told of new skeletons of *Rodhocetus* and a closely related whale called *Artiocetus*.[14] In both cases, there were well-preserved hind limbs that included a particular anklebone, an astragalus, with an es-pecially revealing anatomy (Fig. 12.7). Both ends of this anklebone were rounded with a smooth groove in the middle of the surface, a shape like

13. Gingerich et al., "New Whale."
14. Gingerich et al., "Origin of Whales."

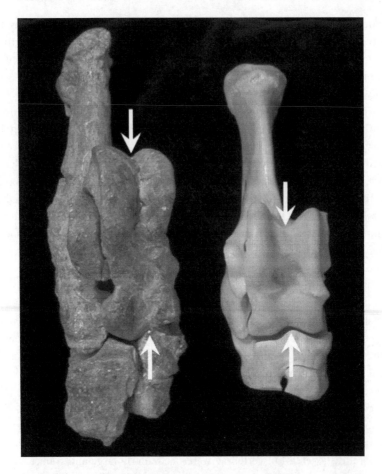

Figure 12.7 — The anklebones of *Rodhocetus* **(left) and a pronghorn antelope,** *Antilocapra* **(right).** The very similar overall anatomy and, especially, the pulley-like form on both ends of the astragalus bone (arrows), unite early whales with *artiodactyls*, the even-toed hoofed mammals. [Photograph by Philip Gingerich, University of Michigan. Used by permission].

the surface of a wheel in a pulley system or for a fan belt in a car. This clear and distinctive pulley-like joint at *both* ends of the astragalus is an anatomical condition not found outside of the lineage of even-toed, land-dwelling, hoofed animals known as artiodactyls (such as deer, cows, hippos, and pigs). It is not present, for example, in even-toed hoofed species, such as horses and rhinoceroses, or in any other mammal. The form of this double-pulley joint limits motion to a forward-and-back movement, which is well-suited to a limb used for running as in the case of artiodactyls. But what's this bone doing in aquatic animals? How fabulous that a specialized bone structure previously considered diagnostic for even-toed hoofed animals on land was found in the fossil remains of two different whales! The double-pulley astragalus has since been discovered in additional ancient whale fossils.[15]

What Does It All Mean?

These discoveries wonderfully illustrate how historical evidence can be used to test predictions generated from evolutionary theory. First, the uncovered fossils strongly confirm the proposed ancestral link between hoofed animals and whales. On the other hand, in an unexpected twist, they suggest that the ancestral lineage of cetaceans came directly from *within* a part of the artiodactyl family tree rather than separately from mesonychids (refer to Fig. 12.4).[16] Although early mesonychids possessed hooves like artiodactyls and early whales, they did not have the double-pulley astragalus that is shared only by the latter animals. This conflicts with the original prediction that cetaceans and artiodactyls developed on different branches from very early wolf-like mesonychids. Of course, what occurred is that we now have new refined models and predictions about the course of evolution; that is, that whales evolved directly from an artiodactyl ancestor.

In total, the new whale fossils unearthed over a 25-year period (even more than I have described here) considerably filled in the gap between land mammals and cetaceans. There are still plenty of questions, to be sure, and there are many speculations as to the complete anatomy and lifestyles of known (and unknown) forms. Nonetheless, each fossil is strikingly consistent with and adds to our previous understandings based on evolutionary theory.

These discoveries have been reinforced by biochemical techniques that allow molecular biologists to directly test the genetic relatedness of organisms (see chapter 14). Comparisons of protein molecules and DNA sequences have solidly linked cetaceans to artiodactyls, particularly to the hippopotamus.[17] A hippo is no whale, but it is certainly intriguing that it is largely an aquatic animal that spends most of its time in water, and it has reduced limbs with hooves. It may sound like science fiction to those who are not accustomed to thinking in an evolutionary sense, but these data strongly suggest that living hippos and cetaceans are closely related and developed along separate evolutionary branches from a relatively recent, shared ancestor.

---

15. e.g., Gingerich et al., "New Protocetid Whale."
16. Boisserie et al., "Position"; Thewissen et al., "Skeletons of Terrestrial Cetaceans."
17. Gatesy et al., "Stability"; Lou, "In Search."

The foreseeable response from the spontaneous creationist community has been mostly to dismiss the new whale fossils on various grounds. These include charging that paleontologists make too much from the fossil remains, maintaining that physiological demands couldn't allow the suggested transitions, and objecting that the discovered species are not direct ancestors to one another.[18] As with other areas in the creation-evolution controversy, spontaneous creationist critiques have been met with reasoned refutations by biologists.[19] If nothing else, the mere presence of hind limbs, hooves, and a double-pulley astragalus bone in whale-like aquatic animals is astonishing evidence for an evolutionary past. These observations are true no matter how extensive, complex, or "obvious" arguments from spontaneous creationists might seem. Modern scientists are interested in deciphering how evolutionary transformations occur, not having to continually return to discussions of whether or not they happened.

Biologists acknowledge that the fossil whale species are not a linear ancestral sequence but they also understand that there is no reason they need to be.[20] A transitional fossil is simply a species that possesses a form or a structure intermediate between those of two others. Specific fossils may be from species on side branches of an evolutionary lineage and still exhibit an intermediate feature shared with other organisms that are in a direct ancestral line. The fact that the intermediate character exists at all is evidence that it probably was present in a direct ancestor. For spontaneous creationists to demand that intermediate fossils must be direct ancestors is an unreasonable, self-imposed stipulation that incorrectly portrays what is required. The argument falsely implies that most fossils will be worthless for testing evolutionary predictions, but that is not the case.

The "big-picture" context of fossil discoveries is often overlooked in debates over transitional forms. For example, *every one* of the fossil whale species occur precisely in sedimentary formations of the predicted time frame between mesonychids and the first fossil cetaceans with modern features. Even further, the overall story in the fossils is a sequential tale in which the oldest are the least whale-like and the more recent are most like modern whales. A noteworthy part of these findings is that the researchers knew in what type of sediments they should look. Some spontaneous creationists claim that scientists desperately conjure up results to match their presuppositions. But if we really consider the personal human elements involved,

18. Camp, "Overselling"; Sherwin, "Scientific Roadblocks."

19. Joiner, "Creationist Mindblocks"; Prothero, *Evolution*, 293–329; Sutera, "Origin of Whales."

20. NCSE, "Transitional Fossils."

the rest of us should readily understand that these hard-earned discoveries are the product of educated and dedicated fieldworkers pursuing the details of a confirmed evolutionary past.[21]

Spontaneous creationists who object to ancient whale fossils seemingly miss or avoid the fact that the fossils fit previously articulated predictions for transitional animals and fill in gaps we were told would never be filled. Most scientists aptly point out that if these findings are not accepted as transitional forms, then it is likely that a person has already determined that intermediate fossils cannot exist. As an evolutionary biologist I am thrilled by new discoveries that enhance our understanding and fill in more pieces of a picture that is consistent with biological evolution. They speak to me of a God who was sovereign over an incredible array of evolutionary transformation through a deep history on earth. But, at the same time, I am saddened that so many of my Christian brothers and sisters cannot share my excitement or see the depth at which biologists have studied such things.

Testing Evolutionary Theory

Beyond what I have already presented, fossils have another incredibly important, but mostly overlooked, value. Since scientists have established the general nature of evolutionary sequences and the timeframes in which they occurred, *all subsequent fossil discoveries represent independent tests of evolutionary theory.* Any find that truly could not be explained under an evolutionary scenario would force a modification of the theory. There have been many surprises, to be sure, in the specifics of particular fossil transitions, but each new fossil has always been consistent with our overall understanding of evolutionary transformations. The fact that new fossils fit predictions so nicely becomes routine for paleontologists because there has been so much verification in the past. Scientists have enormous confidence in what is known about evolutionary progressions. In any debate over biological evolution versus special creation, the reality that almost all fossils fit previous understandings should be viewed as a remarkable confirmation of evolutionary theory!

Any single verifiable find of a bird fossil within an old Devonian rock formation would dramatically falsify the general theory of evolution. Instead, abundant vertebrate remains from this period (415–40 MYA) are limited consistently to fishes and amphibians, while bird fossils appear only in more recent deposits from the Jurassic period on (~150 MYA to

21. See these detailed review articles: Gingerich, "Evolution of Whales"; Thewissen et al., "From Land to Water."

the present). Similarly, real evidence of human existence during the time of dinosaurs would immediately overturn the evolutionary paradigm that pervades all of biology. (Note: some young-earth creationists still report evidence of this sort, even though Christian biologists and young-earth organizations have refuted it.[22]) In the same way, if some unexplainably bizarre fossil were discovered, such as a Carboniferous amphibian with avian wings, it would instantaneously shake the foundation of evolutionary theory. Of course, no such evidence exists and scientists comfortably understand that, after all that has been documented about evolutionary changes through time, no such evidence is likely to be discovered. Instead, new finds of feathered dinosaurs, amphibians with fish gills, and whales with hooves are consistent with evolutionary theory, and each comes from dated fossil sediments exactly where we expect they should be!

22. Hastings, "Rise and Fall"; Morris, "Paluxy River Mystery"

13

Where on Earth?

EVOLUTION IS BEST UNDERSTOOD as a dynamic process that is constantly unfolding. What makes it difficult for most people to visualize is that it is generally an extremely slow progression. This means that what we see today in the world is only a snapshot of where things are at this one moment in time. We can't really watch the sluggish process unfold but we can use today's snapshot to inform us about what has happened in the past.

Biologists understand that a key part of this evolutionary unfolding is events referred to as "speciation." This starts with the isolation of some individuals of a species into a population separated from other members. We can think of foxes on an island isolated from mainland foxes or gophers separated on either side of a major river. On a larger scale, one can imagine toads on either side of an uplifted mountain range or warblers split up and confined to the western and eastern parts of North America by a Pleistocene ice sheet extending down the middle of the continent. In an evolutionary scenario, the individuals of the separated populations will undergo independent modifications over time and often diverge so much from one another that they will become distinctly different biological entities. This means that there are two species where there once was one.

With this view of speciation, evolutionary biologists can make predictions about what we should see in the snapshot of today's organisms around the earth. First, if species are derived as just described, species that are biologically most similar (i.e., they evolved from a common ancestor) should be found in geographic proximity to one another, rather than distributed independently around the globe. In fact, this prediction should generally be true for entire ancestral lineages of species; groups of biologically most similar species should often be found in clusters in different parts of the world. These ideas amount to testable predictions of evolutionary theory. The expectation that biologically similar forms will have close geographical ranges should be most tightly met for species of most recent derivation (i.e., for species pairs that are most similar in traits).

As you can probably guess, observations of animal and plant oc-curence around the earth are generally consistent with these evolutionary predictions. If we choose a single focal animal or plant species and then ask, "Where in the world is the most similar species located?", the most frequent answer is that the species is in the same geographic area. Many illustrations of this pattern will be discussed in the following pages.

An important corollary prediction from an evolutionary perspective is that species will often be *absent* from locations on the earth where its members could successfully live. This absence is expected simply because each species evolves in only one particular geographic region and, typically, its members are unable to move to other places in the world where there are suitable conditions. It follows that where a species is found on the earth will be noticeably correlated with nearness to taxonomic relatives while, at the same time, the species will be missing from locations around the globe that contain perfectly appropriate conditions for survival.

Once again, the pattern of organisms on earth is consistent with this prediction. The fact that many or most species can survive well where they do not naturally occur has been graphically illustrated by widespread plant and animal introductions into new areas. These introductions have occurred due to accidental or deliberate human activities. Here are two extreme ex-amples: (1) Almost all lowland species of birds, reptiles, and mammals on the Hawaiian Islands are foreign species introduced in recent human his-tory from different parts of the world. Ironically, the familiar bird and lizard species most tourists fondly associate with the islands are actually invaders who have displaced the native fauna. (2) The vast majority of wild grassland species throughout the western United States are native to Mediterranean regions on the other side of the globe. Europeans unintentionally brought these species to the Americas with their livestock, and the combined ef-fects of cattle and sheep grazing and the vigor of the introduced grasses have drastically reduced native North American grassland species. It is odd indeed to think that most of the rolling grasslands of the West are largely populated by species from other continents.

In contrast to the previous discussion, we may consider how God might place species around the earth under special creation models. Obvi-ously, we cannot claim to know the mind of God and there is really no way to predict how he might choose to distribute created organisms. However, if God created species independently and placed them in suitable habitats on the earth, there is no particular reason to expect species to be distributed in a way that is consistent with an evolutionary past. Once again, evolutionary theory has explanatory power and provides us with a rational scientific basis for understanding the distribution of organisms around the globe. We can always say that God found pleasure in distributing similar species in groupings within specific regions on the earth, but the more intricately that species' ranges match the predictions of evolution, the less satisfying this suggestion becomes. The distribution of organisms in the world is especially difficult to explain under young-earth conceptualizations that include the notion of a worldwide flood only several thousand years ago. This would mean that animals (or at least all land vertebrates) migrated from Noah's ark after the flood and ended up in worldwide distributions that match the predictions of evolutionary theory.

What Islands Tell Us

The study of biogeography, the distribution of species, is a scientific discipline that blossomed with the natural discoveries of 18th- and 19th-century explorers. Biologists of the time acquired an exploding knowledge about new species from around the world as naturalists aboard sailing ships provided specimens and distributional observations of never-before-described species. Charles Darwin first saw that biogeographical observations were often consistent with the idea that similar species in a specific region evolved from a common ancestral source.

On the voyage of the HMS *Beagle* during the 1830s, Darwin visited both the Cape Verde Islands, situated several hundred miles off the western coast of Africa, and the Galápagos Islands, located a similar distance from the western coast of South America. Each island group contains unique species that are found nowhere else in the world. What struck Darwin most was the species' resemblance to organisms on the nearest mainland. He wrote later in *The Origin of Species*:

> There is considerable degree of resemblance in the volcanic nature of the soil, in climate, height, and size of the islands, between the Galapagos and Cape de Verde Archipelagos; but what an entire and absolute difference in their inhabitants! The inhabitants of the Cape de Verde Islands are related to those of Africa, like those of the Galapagos to America. I believe this

grand fact can receive no sort of explanation on the ordinary view of independent creation.[1]

An evolutionary scenario suggests that some species were able to colonize the volcanic islands from the adjacent mainland and that their descendants changed through time (i.e., they evolved) into new species adapted to the unique available island conditions. The Cape Verde and Galápagos islands are situated at similar latitudes and have similar volcanic island environments. But they are populated not by matching species designed for volcanic islands but by species that most resemble forms from very different environments on the closest mainland. Evolutionary explanations have the power to make sense of these biogeographical observations, while spontaneous creation does not. Why would God create unique sets of species for each of these very small island groups and simultaneously make them appear most similar to species on the adjacent mainland? If the distributions are not the result of an evolutionary past, and they are God's doing instead, that concerning issue of implied deception is raised once more.

Probably the best-described group of species on the Galápagos Islands is a series of birds often referred to as Darwin's Finches. There are fourteen species among the landmasses of the Galápagos chain that all resemble a common finch from the adjacent South American mainland. Various specimens from the mainland and the different islands collected by Darwin were readily grouped together as related finches by taxonomists back in England. What is remarkable within the Galápagos archipelago is the diverse set of bill shapes and feeding behaviors exhibited by the different finch species. Two species of warbler finches have a smallish bill that is used to glean insects from foliage. Another species, the woodpecker finch, has a long and strong bill that it uses to probe for wood-boring insects. Other species are like typical finches and have deep conical bills that are used to crack seeds, but these vary from a small tree finch species with a tiny bill to a very large ground finch species with an oversized beak as big as its head.

Darwin's finches illustrate what biologists call "adaptive radiation," the evolution of many species from a common ancestor to fill multiple available ecological roles (or "niches"). There are no true warblers or woodpeckers on the Galápagos Islands, presumably because none ever made it there from South America. Warbler finches and woodpecker finches have apparently evolved to exploit the island resources normally utilized by species of those types in mainland bird communities.

---

1. Darwin, *Origin*, 398–99.

Species Groupings

The pattern of biologically similar (closely related) species being found near each other can be demonstrated extensively with all sorts of animals and plants. For example, as an avid bird watcher, I can pick out complexes of bird species that occur near each other and seem to have radiated from a shared ancestral lineage. There are four species of non-migratory thrashers (Curve-billed, California, Le Conte's, and Crissal) that occupy arid habitats in the southwestern United States and northern Mexico. All have the same elongated thrasher body form, as well as the stereotypic behavior of thrashing the ground to search for food using their strong feet and long, curved bill. Yet, as distinct species, they differ in coloration and vocalizations and occupy slightly different habitats and ranges. A comparable complex is three species of tanagers (Scarlet, Summer, and Hepatic) with bright red males and yellowish females that fill different habitats during the spring and summer months in the southern and eastern parts of the United States. As bird watchers know, there are other species groupings like these across North America. Evolutionary theory provides a scientific explanation for why the similar species occupy specific geographical regions and occur nowhere else in the world.

Species complexes are also apparent in plants. There are 45 described living species within the plant genus, *Ceanothus*. All distributions are limited to North America, almost entirely in the West, and with particular diversity in California. Many of the species within the complex show considerable overlap in range. They are readily grouped into the same taxonomic genus because they all possess a unique flower structure that unites them (including sepals folded toward the center of the flower alternating with odd scoop-shaped petals — Fig. 13.1). Each of the species is a perennial woody plant, but the species vary from small prostrate ground covers to massive woody shrubs. From an evolutionary viewpoint, the various body forms of the species appear to have been derived by evolutionary processes over geological time in the geographically diverse habitats of the West. Why else should such structurally different plant species from a limited region on the earth share an identical unique flower structure?

A practically endless list of species complexes like these could be provided for various animals from fishes to beetles to rodents and for various plants from wildflowers to trees!

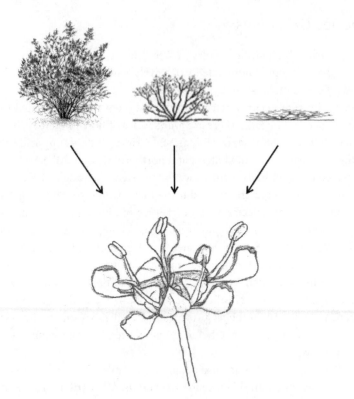

Figure 13.1 — The distinctive flower shared by all *Ceanothus* **species.**
Each petal extends outward like a scoop with a narrow base, the sepals point
inward like arranged triangles, and the stamens stand erect in an expanded
circle. These flower parts vary in color from white to deep purple on *Ceanothus*
species ranging from massive woody shrubs to low spreading ground covers.

The Spontaneous Creationist Response: Microevolution

It may surprise many readers to learn that large numbers of spontaneous
creationists (both young-earth and old-earth adherents) acknowledge that
species complexes like those that I have described are the result of evolutionary processes.[2] They concede that natural selection can modify the
form of plants and animals, even to the extent that they become distinctly
different biological entities. In other words, they accept that new species
with substantial anatomical and physiological changes can develop from a

2. Klotz, "Flora and Fauna."

common ancestral organism. It is worth pausing to point out that this is a major acceptance of biological evolution!

Most spontaneous creationists, however, are quick to clarify that this evolutionary change is limited in extent (e.g., Darwin's finches are still "finches" and all the *Ceanothus* plants are still members of the same genus). It may also be asserted that alterations are limited to within biblical "kinds" created by God. This type of biological evolution is carefully distinguished as "microevolution." This position still rejects "macroevolution," that is, natural processes that have transformed organisms across higher taxonomic levels.

Oddly, for strict young-earth creationists, this should immediately be an unworkable and unacceptable defense. The amount of evolutionary modifications required to produce all of Darwin's finches or 45 species of *Ceanothus* from single common ancestors could not take place in 10,000 to 15,000 years by any known biological mechanisms. No evolutionists would even begin to think that this is enough time for major geological processes to isolate populations and/or for sufficient genetic modifications to accumulate to account for all such diverse species complexes. Creation scientists use the terms, natural selection, genetic recombination, and speciation, as if they are referring to something like the mechanisms studied and explained by evolutionary biologists, but they then extrapolate beyond modern biology to say these can work extremely rapidly to produce huge changes in several thousand years. This is incredibly misleading.

This microevolutionary proposition is particularly unreasonable when it is used to explain the derivation of even larger and more widespread species groupings. For example, leading young-earth flood geologists from AiG and ICR suggest that all living species of cat-like, dog-like, and horse-like species that are *distributed around the world* evolved from single animal pairs released from Noah's ark after the biblical flood less than 10,000 years ago.[3] It is frustrating for scientists and strangely paradoxical to see young-earth creationists misuse evolutionary concepts to explain away positive evidence that supports mainstream evolutionary theory. Let me be clear. Although there are some illustrations of rapid natural selection,[4] there is no real evidence that this can explain the *enormous amounts* of microevolutionary changes accepted by young-earth creation scientists.

On the other hand, it is certainly reasonable for old-earth creationists to use microevolution to account for species complexes because they accept very long time periods over which such processes could operate. But

3. Ashcraft, "Evolution"; Batten, "Ligers and Wholphins"; Purdom and Hodge, "Zonkeys."

4. Catchpoole and Wieland, "Speedy Species Surprise."

the issue then becomes whether there is any real scientific support for the notion that evolutionary change is ultimately limited in extent. No scientist of any persuasion has proposed a mechanism by which genetic modification might be restricted within limits corresponding to biblical kinds or any other boundaries. Spontaneous creationists like to point out that biologists cannot prove that microevolution can be extended to account for macroevolutionary changes and yet, quite significantly, they cannot provide a scientific mechanism to explain why it couldn't. In fact, most evolutionary biologists do not recognize fundamental differences between microevolution within animal and plant genera or families versus macroevolution across higher taxonomic categories. It is very reasonable with our current scientific knowledge to assume that large-scale transformations would be the inevitable result of microevolutionary processes applied over especially long periods. Although it is possible, or even probable, that new additional mechanisms of evolutionary change will be discovered to expand our understanding (see chapter 15), there is no reason to expect that this would alter the fundamental sameness of evolution across all levels, i.e., descent with modification.

The continuity of evolutionary connections across organisms is supported by the fact that every living thing utilizes the same DNA genetic coding to develop body form and control metabolic processes. The genetic instructions of any one organism can be theoretically modified to the instructions of any other organism by conceptually straightforward multiplications, additions, and subtractions of "letters" from the universal DNA alphabet. There are no inherent breaks or discontinuities in the way DNA is structured or used in organisms that imply any limits in possible variations of this instructional language. There is certainly nothing that matches up with biblical "kinds" as defined by young-earth creationists.

Even so, one of the hottest young-earth creationist topics is an attempt to define and delineate the limits of created groups. "Baraminology" begins with the unalterable assumption that distinct categories exist based on specific interpretations of Genesis 1. Investigators have utilized sophisticated terminology and approaches similar to evolutionary biology to described the boundaries of separate biblical groups, called "baramins."[5] For example, all fossil and living horse-like species are aggregated in one baramin, while all fossil and living cat-like species compromise another.

There are two fundamental problems with baraminology that should eliminate it as a scientific proposition (despite the enormous depth of

---

5. Frair, "Baraminology"; Wood, *Animal and Plant*; Wood, "Current Status."

analysis by its followers).[6] First, the notion that there are restricted biological kinds is not based on natural observations. There is nothing that should
incline scientists to focus exclusively on discovering such a thing. In fact,
accumulated scientific evidence as described above suggests exactly the
opposite. Baraminology is clearly a pursuit to force the natural world to
demonstrate what young-earth creationists already believe from the biblical
text. The second, and most convicting, challenge is that proponents have not
provided a scientific rationale or any biological mechanisms for why or how
the boundaries of baramins should exist or be maintained in nature. I urge
biblical Christians to understand that this strict adherence to biblical kinds
is not an open-minded investigation of God's creation.

Our common experience with living organisms inclines us to believe
that creatures occur in distinct groupings, such as cats versus dogs, hoofed
animals versus whales, and reptiles versus birds. But when a totality of biological information is considered, these distinctions fade away. This is just
what evolutionary theory predicts. Previous chapters have discussed evidence for evolutionary connections not just among similar species but also
across major organism groupings. These include homologous forelimbs that
unite all amphibians, reptiles, birds, and mammals; transitional fossils that
dissolve the boundaries between fish and amphibians; and developmental
peculiarities and fossil discoveries that link deer, cows, hippos, whales, and
dolphins. Even further, there are significant biogeographical observations
that suggest some very large biological groupings came about by evolutionary processes. Please read on.

Biogeography on a Larger Scale

Birds are among the most mobile creatures on earth and it is not surprising
that many higher-level bird categories are represented by species distributed
around the world. Familiar examples include ducks, herons, gulls, pigeons,
owls, and woodpeckers. On the other hand, many bird types are limited to
particular regions of the globe as would be expected if some evolutionary
lineages arose in restricted geographic areas. This observation is especially
true among the songbirds, which ornithologists have good reason to believe
are the most recently evolved species.

For example, many families of songbirds contain species that occur exclusively in either the New World (North, Central, and/or South America) or
the Old World (Europe, Africa, and/or Asia). Table 13.1 lists some songbird

6. Elphick, "Baraminology"; Gishlick, "Baraminology."

families that follow these patterns.[7] Each of these bird groups contains a diverse array of numerous species that occupy many types of habitats, but only within one or the other of the global regions. For example, there are 283 species of tanagers and 241 species of ovenbirds in the New World, but not a single species of either type in the Old World. In contrast, ornithologists have identified 310 species of babblers and 139 species of bulbuls in the Old World and not a one in New World areas.

An evolutionary view suggests that while mechanisms of speciation were occurring within the different continental regions, birds among these groups did not disperse across barriers between the New and Old Worlds. As already stated, any distribution of species can be explained simply as the way God chose to place different bird forms on earth. But it is also true that the pattern of geographic distributions described here is consistent with the evolutionary derivation of numerous related species within effectively separated Old and New World regions.

Table 13.1 — Example songbird groups with species distributions limited to either the New World (North, Central, and/or South America) or the Old World (Europe, Africa, and/or Asia).

Bird Group	Taxonomic Family	Number New World Species	Number Old World Species
Tyrant Flycatchers	Tyrannidae	432	0
Tanagers	Thraupidae	283	0
Ovenbirds	Furnariidae	241	0
Antbirds	Thamnophilidae	213	0
Wood Warblers	Parulidae	116	0
Wrens *	Troglodytidae	86	1

7. Del Hoyo et al., *Birds of the World.*

Bird Group	Taxonomic Family	Number New World Species	Number Old World Species
Old World Flycatchers	Muscicapidae	0	117
Starlings **	Sturnidae	0	112
Babblers	Timaliidae	0	310
Bulbuls	Pycnonotidae	0	139
Weaver Finches	Ploceidae	0	115
Larks ***	Alaudidae	1	96

* 1 species native in Old World
** 2 species introduced to North America
*** 1 species native in New World

An evolutionary scenario is more fully supported by two additional observations. First, biologists often notice that New and Old Worlds regions contain what are called "ecological counterparts." For example, "flycatchers" are birds that sit in wait on a relatively exposed perch, fly out to catch insect prey, and then return to their resting location. This flycatcher role in nature ("niche") is occupied by various species in just about every habitat throughout the world. However, the role is filled in different environments from Canada to the tip of South America by over 400 species from a single New World bird group called the "tyrant flycatchers" (Family Tyrannidae). In bird communities of Europe, Asia, and Africa, the flycatcher niche is occupied by taxonomically different species, particularly by over 115 species of "Old World flycatchers" (Family Muscicapidae). Many of the species on opposite sides of the Atlantic Ocean appear to be perfectly interchangeable in terms of their behavior and the habitats they occupy. In this light, their geographic distribution seems to tell a story of an evolutionary past.

The second observation, which is especially consistent with evolutionary prediction, is that very many bird species have been artificially introduced to areas quite distant from their native range and have survived perfectly well (often to the detriment of native birds in the new area). This does not fit well with the notion that God specially designed species and individually placed each one in a specifically chosen location on the earth.

Large-scale evolutionary patterns can even be observed for *entire communities* of organisms. A relatively narrow expanse of land separates coral reef communities on the west and east sides of Central America. However, the aquatic invertebrate and fish species on either side of the Central American landmass are far less similar than one might guess. In fact, the west coast marine communities contain many species that show the closest biological affinities with invertebrate and fish species from all around the Pacific Ocean, rather than with the species just several miles away on the eastern side of Central America. Likewise, the east coast reef assemblages contain many species that show distinct Atlantic Ocean connections. These observations are what we expect if membership in these communities reflects the evolution of species in separate west and east coast aquatic environments after the formation of a Panamanian land bridge.

Drifting Continents

Even though the biogeography of species has generally provided support to evolutionary theory, there have been some historically difficult-to-explain, even embarrassing, exceptions. For example, species of a specific group of small freshwater fish, called cichlids, are found only in widely distant rivers of South America and Africa. How could these cichlid species have evolved from a common ancestor when they occur on widely separate continents? Similarly, why are flightless, ostrich-like birds, called ratites, found only in the separated regions of Australia (emus), New Guinea (cassowaries), New Zealand (kiwis), and the southern reaches of South America (rheas) and Africa (ostriches)? And why are fossil forms of camels and elephants found not only in Europe, Asia, and northern Africa, but also in faraway North America? Biologists had proposed such things as the independent evolution of each flightless bird group, long-distance travel over temporary land bridges for large mammals, or they simply acknowledged they did not have a good explanation for such observations (e.g., cichlids).

It is probably an instructive digression to again point out that these problematic cases did not lead scientists to reject evolutionary theory. The weight of evidence still pointed to an evolutionary view, so biologists proceeded with the assumption that all biogeographical observations would have some sort of evolutionary explanation. In response, spontaneous creationists can claim that scientists have an evolutionary bias that will not let them consider a special creation alternative. But this situation actually helps demonstrate further how evolution is a major *theory* in science, not simply a hypothesis. A few hard-to-explain observations ought to be fatal for a weakly supported hypothesis, but they should have little effect on a unifying and useful theory that has been well-verified by previous observation and experimentation.

In a wonderful example of the development of a new idea in science, the problematic cases mentioned above, and others like them, were explained once scientists gained an understanding of the phenomenon known as continental drift or "plate tectonics." A German scientist named Alfred Wegener originally proposed in 1915 that large continental landmasses could move over the earth surface. The idea received little support because there was no known mechanism for such a thing and no real confirmation. A slow accumulation of evidence convinced scientists to take a closer look and we now have a modern geological understanding that continental plates do indeed float on underlying basalt. Knowledge of continental drift mechanisms now helps scientists explain mountain uplifting, earthquakes, and volcanic activity that occur where two plates collide, slide past each other, or spread apart.

Geologists have been able to piece together a well-accepted sequence of where the continental plates have been positioned on the earth's surface through time. They use such things as the position of shared rock formations on different continents and the orientation of magnetic particles deposited on continents at different times. In summary form, it is suggested that all the continental landmasses drifted toward one another during the Permian geological period to form one large continent called Pangaea (Fig. 13.2a). After this, Pangaea fragmented into two major continents, a northern Laurasia and a southern Gondwana (Fig. 13.2b). Further movement led to the breakup of Laurasia into North America and Euroasia. At the same time, Gondwana split into the separate landmasses of South American, Africa, Antarctica, India, and Australia. Finally, the separate continental plates drifted to their current New World and Old World positions (Fig. 13.2c).

Plate tectonics provided missing pages to the story of life on earth and helped explain biogeographical observations that had confounded biologists. For example, cichlid fish occur in separated regions of South America and Africa because their ancestors swam in the freshwater rivers on the southern continent of Gondwana. Flightless ratite birds occur in Australia and the southern reaches of South America and Africa because these regions were once joined in the Gondwana landmass when the ancestors of these bird types first evolved. Fossil camels and elephants occur in North America, Europe, Asia, and Africa because there were major connections among all these continents during the Tertiary period when these fossil species existed.

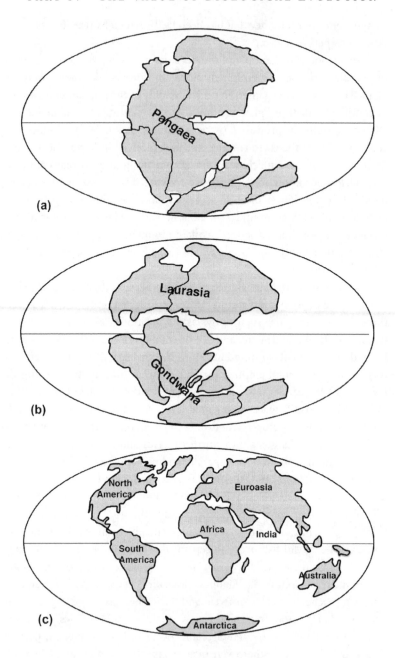

Figure 13.2 — The position of continents through time. (a) The supercontinent of Pangaea around 225 million years ago. (b) The two major continents of Laurasia and Gondwana around 140 million years ago. (c) The continents today.

With an understanding of continental drift, the picture of past distributions of animals and plants has become increasingly clear. For one final example, distinctive fossils species of marsupials, large reptiles, and seed-bearing trees have been discovered on the frozen Antarctic continent and they match fossils that are found in the southern regions of Africa and South America. These findings make sense if Antarctica was once positioned much further north in the combined Gondwana landmass along with Africa and South America. This is really a rather remarkable confirmation of continental drift and evolutionary theory, since no marsupials, reptiles or trees could possibly have lived on Antarctica if it was always located where it is today.

Conclusions

The illustrations I have given in this chapter are just a sample of abundantly recurring patterns of animal and plant distribution. An evolutionary scenario provides an explanation for what we see in the natural world while special creation positions do not. There is no reason to *expect* that God would distribute species in a way that so fully correlates with evolutionary relatedness. For those who still hold to a global flood, there is also no reason for animals to have migrated to locations around the world to imply these patterns. For Christians, is it not more reasonable that God's creation is telling us something real about its evolutionary history?

14

The New Frontier

As a biologist most deeply trained in whole organism biology and ecology, I playfully kid my colleagues that biological topics involving anything smaller than a cell are not as interesting as my fields of study. In reality, I excitedly acknowledge that the most fascinating advances in recent years have been in the ability of scientists to understand and manipulate the very molecules that are responsible for the activities of living things. Astounding progress in modern techniques now allow scientists to not only reveal the precise structure of individual DNA and protein molecules, but to also build and alter these molecules. Biologists increasingly understand intricate aspects of cell operations and they creatively manipulate how cells function. One product of modern research has been delineating the specific DNA code for a variety of organisms. In fact, after thirteen years of intensive work, a coordinated, worldwide Human Genome Project completed this amazing feat for humans in 2003. Gene sequences have been mapped for all the chromosomes in human cells and this information is now readily available on the Internet. This remarkable achievement was merely formulated in the dreams of scientists as little as forty years ago.

One artifact of biochemical advances is that evolutionary scientists have been given new methods to test the ancestral relationships among organisms. Individuals within a single species share very similar DNA instructions and mostly use the same molecules for their metabolism. This genetic uniformity is generally maintained as DNA is reshuffled in the sexual reproduction of new offspring. However, evolutionary alterations necessarily involve a shift in a population's genetic makeup due to accumulated modifications in DNA. We understand that greater evolutionary change over time should be accomplished through generally greater changes in the structure of DNA, as well as in the form of the molecules that are made from the DNA instructions. This leads to very clear predictions. In particular, species judged to be closely related using other lines of evolutionary evidence should show more similarity in their DNA and constructed molecules than species understood to be more distantly related. In fact, we can predict that the *number or degree of differences in various molecules* among species should correlate with what has been previously proposed as the *evolutionary distance* between species.

These molecular predictions amount to significant independent tests of evolutionary theory. In other words, these are brand-new approaches for investigating the validity of previously described evolutionary relationships. Spontaneous creationists have often charged that evolutionary scientists simply piece together evidence from various aspects of biology. But the biochemical techniques now used to make molecular comparisons have arrived after the major framework of evolutionary theory was developed. This means that tests of evolutionary predictions based on molecular data cannot be dismissed in the same way. It would seem that the results should be allowed to have a notable impact in support or rejection of biological evolution.

The Basics of Molecular Genetics

I'm sure that a short overview of cell operations will be helpful for some readers. By the mid-1950s, scientists had discovered that DNA molecules contain the code that determines each organism's structure and functioning. It is now understood that most functional stretches of DNA, called *genes*, hold specific information for the construction of *protein molecules*. These proteins, in turn, are typically enzymes or building blocks that ultimately determine an organism's attributes.

DNA molecules have a form described as a "double helix," which is something like a ladder twisted into a spiral. Each step in the ladder is composed of important units called *nucleotide bases*. These nucleotides are limited to only four possible options: adenine (A), guanine (G), thymine (T), and cytosine (C). One of the startlingly simple realities of molecular genetics is that the information carried on the DNA of all organisms is determined by the specific linear sequences of these four bases along the ladder.

All proteins, in turn, are composed of a linked series of twenty different possible units called *amino acids*. Proteins are like long chains of pearls in which there can be varied sequences of any combination of twenty different pearl types. In general, when a specific gene segment on DNA is activated, the specific nucleotide base sequence of the gene determines the precise sequence of amino acids in a constructed protein. The crucial point is that the sequence of DNA bases directs the sequence of amino acids in a protein, and this dictates which specific type of protein is made.

There are, of course, many more details in DNA activation and protein construction, but this overview should provide an adequate framework for understanding the discussions that follow. The relationship of DNA to proteins described here is a universal principle in all living things on earth. In an extraordinary statement about the unity of life, scientists have found that this DNA-to-protein principle is the same from the simplest bacteria and protozoans, to earthworms, frogs and fishes, to the tallest trees and humans.

Comparing Proteins

As a protein is put together in a cell, the chain of amino acids folds upon itself to take on a unique three-dimensional form that determines the protein's characteristics. This form depends specifically on the total number and sequence of amino acid types within the molecule. Some positions in the sequence are critical and must be precisely the correct amino acid out of twenty possible options for the protein to function well. However, a large number of positions in most proteins can be occupied by a variety of alternative amino acids without significantly altering the effectiveness of the molecule. These amino acids are essentially filling a location within the protein structure without directly influencing the molecule's behavior.

Any DNA mistake (mutation) that alters an amino acid at a functionally important position will generally produce a compromised or dysfunctional protein. An individual organism with a mutation like this would typically not be expected to survive. However, random mutations that substitute amino acids at less important "spacer" positions do not harm the protein and do not affect the survival of individuals. These amino acid changes can be retained and passed on to offspring. This means alterations at the functionally unimportant amino acid positions can accumulate through a long lineage of ancestral organisms. Given this understanding, it is possible to make a significant evolutionary prediction: *species that are proposed to be evolutionarily more distant from one another should have accumulated more random changes in amino acids than species with closer ancestral connections.*

This prediction has been repeatedly tested through the comparison of many types of protein molecules across different organisms. One early and famous example is the study of what is known as cytochrome-c. This molecule is involved in energy metabolism within almost all living organisms, from single-celled yeast to humans. In some remarkable and fascinating research, it has been shown that cytochrome-c molecules taken from all kinds of different organisms (including humans, pigeons, horses, rats, tuna fish, and flies) will operate successfully inside yeast cells that have been experimentally stripped of their own cytochrome-c molecules.[1] This is true even though all the molecules have *differing amino acid sequences*. In other words, the precise sequence of amino acids within cytochrome-c differs from one organism to another but does not damage the operation of the protein.

Molecular biologists have determined the amino acid pattern in cytochrome-c for a large variety of organisms.[2] The results show a general match between species amino acid differences and evolutionary relationships previously determined from other criteria. This can be illustrated by comparing the 100 amino acid sequence of cytochrome-c in humans with the same protein in other species (see Fig. 14.1). The cytochrome-c molecules of chimpanzees are identical to humans, while Rhesus monkeys differ by one amino acid. Other mammals such as rabbits, pigs, cows, whales, dogs, and horses show differences of nine to eleven amino acids. Reptiles, amphibians, fish, and insects differ by progressively greater numbers of amino acids corresponding to the predicted evolutionary distances from humans to these groups. Alignment with evolutionary predictions is not absolutely perfect (i.e., the kangaroo should be more distant than other mammals and the duck, like the other birds, is expected to show more amino acid differences than all the mammals), but the overall correspondence is remarkably good.

It is important to remember that these are comparisons within a cytochrome-c molecule that functions in the same way across all organisms. There is nothing in terms of the molecule's operation within the various species that directly explains amino acid variations and yet the differences match predictions based on evolutionary relationships. The results are just what are expected if organisms share a common ancestry and occasional mistakes in genetic information are transmitted from generation to generation.

1. Theobold, "29+ Evidences," Part 4.1.
2. Gray, "Biochemistry and Evolution," 265–74.

human	
chimpanzee	**0**
rhesus monkey	**- 1**
rabbit	--------- **9**
kangaroo	---------- **10**
pig and sheep	---------- **10**
cow	---------- **10**
gray whale	---------- **10**
dog	----------- **11**
horse	----------- **11**
duck	----------- **11**
chicken	------------- **13**
turkey	------------- **13**
penguin	------------- **13**
rattlesnake	-------------- **14**
snapping turtle	---------------- **15**
bullfrog	------------------ **18**
tuna fish	-------------------- **21**
screwworm fly	---------------------------- **27**
silkworm moth	------------------------------- **31**
wheat	-- **43**
baker's yeast	--- **45**

Figure 14.1 — The number of amino acid differences in cytochrome-c molecules between humans and other species. The amino acid sequence of humans is the reference point to which other species are compared. Numbers indicate how many amino acid positions differ in each species compared to the human sequence.

Straight to the DNA

Biologists have also compared species for differences within specific genes on the DNA itself. I mentioned in chapter 12 that molecular studies pointed to the hippopotamus as the closest-living relative of whales and dolphins. The earliest reports were comparisons of DNA nucleotide sequences. One study examined a gene for fibrinogen (an essential protein in blood clotting) and another looked at two genes for casein milk proteins (molecules contained in the milk of female mammals).[3] Assessment of each of the three genes yielded the results depicted in Figure 14.2.

This branching evolutionary tree diagram, called a cladogram, is constructed by analyzing the number of nucleotide differences among *all* species parings. Species that are most similar in their DNA sequences are closest together. For example, whale and dolphin species have DNA that

3. Gatesy, "More Support"; Gatesy et al., "Evidence."

is very alike in nucleotide pattern, so they are shown as closely connected. Whales and dolphins were also found to have DNA more similar to the hippopotamus than to any other hoofed mammal, so the hippo is the next-closest connection. The diagram indicates a shared whale-dolphin-hippopotamus ancestor in the past at the fork that unites these forms. The rest of the cladogram can be read in similar fashion. For example, cow and sheep DNA is most like that of deer species and the next-closest hoofed animal DNA is from giraffes. Horses have the largest differences in DNA sequences from all other species and the cladogram illustrates that they are most distantly related in the past from the other hoofed mammals.

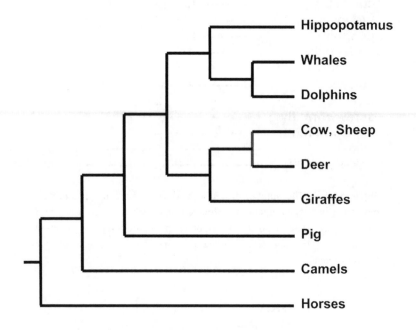

Figure 14.2 — An evolutionary tree (cladogram) for whales, dolphins, and hoofed mammals based on the number of difference in DNA nucleotides in genes for fibrinogen and casein milk proteins. Species that are close together in the diagram are the most similar in DNA sequences. See the text for further explanation.

The ancestral relationships revealed in the Figure 14.2 agree with previously understood associations among hoofed mammals in most respects. There were some surprises. For example, pigs had been thought to be more distant than camels from most of the other forms. This is an interesting, but not a major revelation. Of course, the most intriguing disclosure was that whales and dolphins fall right in the middle of the other artiodactyl species, and they are most similar to the hippopotamus. We see that whales and dolphins are also more similar in DNA to cows, deer, and giraffes than they are to pigs, camels, or horses. This information conflicts with past proposals that cetaceans are only distantly related to hoofed animals through a common mesonychid ancestor (see Fig. 12.4). The results are consistent, instead, with the conclusion that whales evolved from a more recent ancestor within the artiodactyls that had a double-pulley astragalus bone!

These same species have been studied through comparisons of many other DNA segments. One project produced a report comparing DNA base sequences within seventeen different genes.[4] The results from the different genes did not line up in all respects and it is important to acknowledge that there is some variance in molecular data. Even so, the close connection of whales, dolphins, and the hippopotamus was confirmed in the majority of cases. Also, data from all genes showed cetaceans falling within the produced cladograms, and they never once suggested that whales and dolphins were distant cousins separate from living hoofed mammals.

The most common spontaneous creationist critique of molecular comparisons is to focus on anomalies and to emphasize that exceptions somehow invalidate the entire process. But this approach provides an incomplete picture that overlooks or ignores the rather remarkable finding that data for the vast majority of species coincide with evolutionary predictions. Others have rejected molecular studies by presenting a false impression of what evolutionary theory predicts. A particular mistake is misunderstanding the measure of evolutionary distances on an evolutionary tree. Data are used to construct total distances between *living* species at the tips of lineage branches by tracing through ancestors at the forks of a cladogram (e.g., a difference of sixteen amino acids between deer and giraffes would be assumed to represent eight differences from each back to the shared deer-giraffe ancestor). Cladograms are not based on noting differences between a living species and a distant ancestor for which no data exist.[5]

The findings described here provide powerful support for the general idea of biological evolution. Comparable results have been produced using numerous other proteins and gene sequences in the study of many different types of organisms. The number of similar studies have increased on a regular basis. The results primarily confirm previously determined evolutionary relationships (as well as answer really interesting questions). The fact that many of the studies are based on variations in molecule structure that do not change the molecule's function is especially significant.

If God created all species independently, he could have indiscriminately supplied organisms with different forms of cytochrome-c or other

4. Gatesy et al., "Stability."

5. For a more complete explanation of this argument and its difficulties, see Gray, "Biochemistry and Evolution," 271–74.

proteins without having any effect on the biology of species. Why should the amino acid sequences of cytochrome-c match the evolutionarily relatedness of species when any number of different versions of this molecule would work perfectly well in every species? No argument can be made that mammals share similar molecules because they must have a "mammalian" form of it. There also can be no arguments that cytochrome-c differences relate to the mode of life of different species. For example, flying bats have cytochrome-c molecules that are more similar to dolphins and monkeys than they are to flying birds. Likewise, soil-dwelling millipedes are more similar to crabs and butterflies that are related through an evolutionary past than they are to earthworms that shares the same soil habitat. Biological evolution provides the explanatory power to make sense of these biochemical observations.

DNA without Function?

One of the great surprises from the detailed sequencing of DNA was the initial discovery that a fair amount of each organism's genome did not appear to be organized into genes that code for operational proteins or to accomplish other known functions. In other words, it looked liked there were many stretches of nonfunctional base sequences. This seemingly functionless DNA has been explained as the accumulation of genetic accidents in the long ancestral history of organisms (such as DNA duplications and random insertions, both of which are well-studied in genetic research). Paradoxically, the presence of DNA like this is understandable in one sense. This is because there appears to be no selective pressure and, importantly, no known cellular mechanism to eliminate DNA that has no impact on the biology of an organism. This means that many past and present "glitches" in the construction of DNA molecules could be simply carried into the future.

Before I go on, I need to point out that most scientists understood the irony of designating any part of the DNA as "functionless." On the face of it, there seems to be something incongruous in the finding that the all-important genetic and hereditary material in our cells contains accumulated garbage that we carry around and pass on to our offspring for no required purpose. In reality, geneticists are investigating ways in which a lot of poorly understood sections of DNA are relevant in the overall operation of cells. Hundreds of scientists from around the globe have collaborated for over ten years in a coordinated effort called the ENCODE project.[6] This research has shown that many sections of non-coding DNA actually function as promoters or suppressors of protein-coding genes, and other functional attributes

---

6. ENCODE Project Consortium.

are being discussed and discovered. Our latest understanding is that much of the DNA works to supervise or manage the actions of the genes that produce proteins.

At the same time, it is absolutely incorrect for spontaneous creationists to argue that all of the currently designated functionless DNA will be shown to have operational roles in the future. This is because some of it is clearly distinguishable as damaged versions of known working genes.

Vestigial Genes

One of the clearest examples of nonfunctional DNA is what geneticists call vestigial genes. A good illustration involves a gene required for the production of ascorbic acid in mammals.[7] Most of us know this molecule by its alternate designation of "vitamin C." The majority of mammals have a useful protein-coding gene for an enzyme (L-gulonolactone oxidase or "GULO") that allows them to synthesize their own ascorbic acid molecules. However, almost all primates (i.e., tarsiers, all monkeys, the great apes, and humans) lack the ability to make ascorbic acid. They must instead obtain it in the food they ingest (this is why many humans use vitamin C supplements to ensure its intake). It may seem logical to assume that since most primates don't manufacture ascorbic acid they simply lack the gene for making it. This is not the case! Instead, the DNA of the deficient primates contains a nonfunctional mutant copy of the GULO gene. This dysfunctional DNA sequence is clearly recognizable as the gene for ascorbic acid synthesis, due to the overall matching pattern of DNA bases, but there are noticeable errors in the sequence that interfere with normal production of the GULO enzyme.

Why should tarsiers, all monkeys, and great apes contain a gene that does not operate correctly that also matches a functional gene found in other mammals? The best explanation is that the GULO gene was disabled by mutations in a common ancestor to all these primates and that the dysfunctional gene has been passed on ever since to its descendants (see Fig. 14.3a). This observation parallels remnant vestigial structures seen in some animals, such as useless eye structures in blind moles, hind limb buds in some snakes, and dangling side toes in the feet of elk (see chapter 11). The nonfunctional GULO gene of tarsiers, monkeys, apes, and humans is a *vestigial gene* that can be explained as a remnant of a previous ancestral condition when the gene was operational.[8]

7. Max, "Plagiarized Errors," Part 2.2; Max, "A Response"; Ohta and Nishikimi, "Random Nucleotide."

8. Venema and Falk, "Signature in the Pseudogenes."

These primate species have long been thought to share a common ancestry that separated them from the lineages of lemurs and other mammals. The vestigial GULO sequence provides further confirmation for that conclusion. If God independently created mammalian species, there is no reason to expect a nonfunctional version of the GULO gene in *any* species. There is certainly no reason why it should be confined to a primate lineage consistent with previously determined evolutionary relationships.[9]

Duplicated Genes

A second category of nonfunctional DNA is the product of events called gene *duplications*. Geneticists have found that it is fairly common for recognizable sequences of DNA to be repeated along the length of a DNA molecule. These repetitions mean that there are often multiple copies of the same gene. For example, several copies of the gene for hemoglobin proteins occur as a "cluster" of genes on the same human chromosome. The repeated genes can be exactly the same as one another or they may vary somewhat. If the genes vary only a little in nucleotide sequence, they may lead to slightly different versions of the protein product. Another option is that a copy of a repeated gene can be a disabled DNA segment; this is a nonfunctional pseudogene (see Fig. 14.4).

This kind of pseudogene closely resembles a known working gene but it is not operational because an important nucleotide base is altered, bases are missing or extra bases have been added compared to the normal gene sequence. The nucleotide changes interfere with the DNA code and this blocks the pseudogene's ability to yield a protein. A pseudogene like this does not harm the individual organism in which it is found because functional copies of the gene are still present to make the needed protein product. The disabled gene is retained in the genome of an organism and it is passed on to its offspring. With the pseudogene as a nonfunctional passenger, there are no constraints on additional mutations in nucleotide bases and these should accumulate steadily through ancestral lineages without consequence to individual organisms.

---

9. Challenges to these GULO conclusions are addressed in Venema, "A Tale."

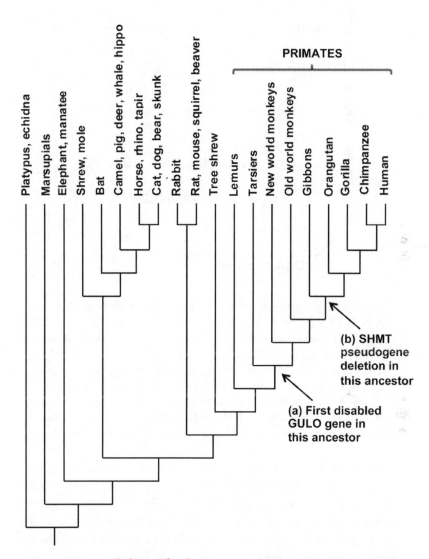

Figure 14.3 — A cladogram for the primary mammal groups and primate sub-groups. (a) All primates other than lemurs have errors in the nucleotide sequence of the GULO gene that cause the gene to be nonfunctional. This suggests that the gene was disabled in the common ancestor to these mammals. (b) Great apes and humans have an identical 11-nucleotide deletion in their SHMT pseudogene. This suggests that the deletion event occurred in the common ancestor of great apes and humans.

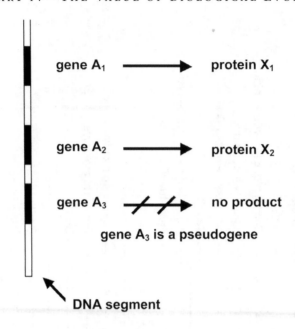

Figure 14.4 — A pseudogene that arises from a duplicated gene. Multiple copies of the same gene can be found together on one stretch of a DNA molecule. In this illustration, three copies of gene A are recognizable due to the overall high similarity in their nucleotide bases. Versions A_1 and A_2 differ enough in their nucleotides that they produce slightly different versions of the same protein, X_1 and X_2. However, gene version A_3 has nucleotide alterations, deletions or additions that render it nonfunctional, and it does not lead to any protein product. Gene A_3 is a disabled pseudogene that will accumulate random alterations in its nucleotides as it is passed on from generation to generation.

One known duplicated pseudogene is around 2,000 nucleotide bases long, and it is recognizable as a marred version of a gene for a specific enzyme (SHMT).[10] It is easily located in the genome of all primate species, from lemurs and monkeys to great apes. One thing that stands out in studies of this pseudogene is that all species of the "great ape" subgroup (including gibbons, orangutans, chimpanzees, gorillas, and humans) are missing the *exact same* sequence of eleven side-by-side nucleotides out of the middle of the gene. The SHMT pseudogene in other primate species is not missing this distinctive set of eleven bases.

How can we account for such a specific structural artifact in the middle of a functionless gene only in great apes? An evolutionary explanation is that an eleven-base deletion occurred in the common ancestor of these

10. Devor et al., "A Unique Genetic Marker."

species and that all living descendants have inherited this condition (see Fig. 14.3b). Great apes have long been classified by the use of other evidence as a united lineage that diverged from other primates. Once more we are faced with molecular data that is consistent with evolutionary predictions and makes little sense if God specially created these organisms.

The studies I have described in this section are by no means unique. In fact, there are now hundreds of published research endeavors utilizing vestigial genes, duplicated pseudogenes, and other types of non-coding DNA (e.g., transposons, retroposons) that have helped scientists elucidate evolutionary relationships among organisms.[11] They have added powerful support for evolutionary theory.

Human Affinities

I finish this chapter by venturing into the topic I have not yet addressed directly but that I am well aware gives evangelical Christians and many others particular difficulty. It is one thing to discuss evolutionary relationships among non-human organisms but there is something that makes a lot of people uncomfortable when humans are included in the discussion. I will approach the unique characteristics of humans and our spiritual nature, especially in the context of biblical passages, in chapters 16–19.

With that said, what do molecular investigations suggest about the relationship of humans to other organisms? Much of this chapter has pointed to the fact that humans fit very well into comparative studies of animal species, and numerous lines of evidence suggest a very close relationship of humans to primates, especially to the great apes and particularly to chimpanzees. I will simply state what some of the data are: The amino acid sequences of human and chimp cytochrome-c molecules are identical. Comparisons of additional proteins and examination of specific DNA genes always show that humans are more similar to chimpanzees and gorillas than to any other mammals. Results routinely show that amino acid and nucleotide sequences are either identical or differ only in a minor way.[12] Humans, chimpanzees, and gorillas also share various identical pseudogene sequences that are not present in any other species. Finally, one distinctive piece of nonfunctional DNA (a transposon) is found on the Y (male) chromosome only in humans and chimps but not gorillas or other primates.[13]

11. Shimamura et al., "Molecular Evidence"; Theobold, "29+ Evidences," Part 4.3.
12. Venema, "Genesis and the Genome."
13. Wimmer, "Direct Evidence."

Much has been written about the apparent percentage difference between human DNA and the DNA of other organisms, especially primates. Readers may have heard that human and chimp DNA is anywhere from 95% to 99% the same. These numbers have changed as various methods have been utilized and as researchers have calculated the values using different criteria. Complete sequencing of the chimpanzee genome was accomplished by an international team of 67 scientists and reported near the end of 2005.[14] Direct comparisons to human DNA at protein-coding genes show more than 98.5% similarity. These are fascinating data and the comparison of human and chimp DNA will most certainly lead to a better understanding of our own biology (including diseases). The significant message is that no matter what biochemical method is used, humans have always been found to be most similar to chimps and gorillas, less similar to other primates, and progressively less similar to other mammals, just as evolutionary biologists have long predicted.

14. Chimpanzee Sequencing and Analysis Consortium, "Initial Sequence"; Culotta, "Chimp Genome."

15

How Did That Happen?

I HAVE OFFERED A sample of natural evidence in the last four chapters to illustrate why biologists are persuaded to an evolutionary view. A particularly compelling aspect of this evidence is that so many varied kinds of biological information point to the same conclusion. The observations demonstrate that virtually everywhere we look within the study of living things we are confronted with an evolutionary scenario. When biologists say they see indications of evolution throughout biology, they really mean it. Only evolutionary theory provides an explanatory framework for making sense of the totality of information.

One evolutionary story within these pages demonstrates how multiple lines of evidence can bring us to an especially compelling conclusion: the proposition that whales evolved from land-based hoofed mammals. Supporting observations include (1) odd anatomical structures in adult whales (e.g., vestigial hip bones), (2) peculiarities in developing whale embryos (e.g., migrating nostrils, temporary hair and teeth), (3) fossil forms with transitional features (e.g., whale-like species with hind limbs and hooves), and (4) very similar DNA and protein molecules among living whales and ungulates. Biologists see these observations like a bunch of scattered signposts pointing to the same location. The idea that whales evolved from a hoofed ancestor is not only supported but we begin to understand that it is strongly confirmed.

The journey to this point has considered the "what" of evolution, that is, the general concept of evolutionary change. But what about the "how" of evolution? What remains is a discussion of the possible *mechanisms* of evolutionary modifications. Can biologists provide specifics about the steps and processes involved?

It is important to pause and consider that the evidence presented thus far is, by itself, strong "proof" for biological evolution. In other words, it is not necessary to have complete knowledge of how evolution happened for there to be a convincing framework that it did indeed take place. These distinctions are significant because arguments from spontaneous creationists commonly confuse them. Critics point to questions and disagreements among biologists about evolutionary mechanisms and claim that this seriously undermines the entirety of evolutionary theory. Others say that unless scientists can explain precisely how evolution happened, or demonstrate it in a lab, then there really is no evidence at all. This was one of the favorite approaches of ID leader Philip Johnson, as he would object, "Evolutionists point to the fact of evolution but can't provide an explanation of the means."[1] This rationale is both inaccurate and misleading.

The logic would be the same if archaeologists found convincing evidence that humans inhabited an island very distant from any other landmass, such as ruins of buildings and pottery shards, but critics refusing to believe any persons ever lived at that location because scientists couldn't say precisely by what method and by what route humans traveled there. Not knowing the method or route of travel, or observing scientists debate about possibilities, does not negate the fact that humans were once present. Natural evidence solidly demonstrates that evolutionary transformations occurred over the past, just as building ruins and pottery pieces clearly indicate that people once lived on an island. In the same way, uncertainties about specific mechanisms of evolutionary change do not negate or constitute difficulty for the general concept of evolution.

Nature Making Choices

Biologists do not have a full picture of the processes of modification over evolutionary time, especially during the earliest periods when only single-celled creatures existed. Even so, we do understand two mechanisms that most certainly played important roles. These are, *natural selection* (the favored survival of individuals who possess the best-suited traits) and the *appearance of new genetic options* (favorable mutations). It may surprise

1. Johnson, *Darwin On Trial*, 10.

some readers that essentially all participants in creation-evolution debates accept these currently known mechanisms. This includes both young-earth and old-earth creationist groups, as well as intelligent design advocates. The heart of the issue is exactly how these mechanisms operate and to what extent they can modify organisms over time.

A few clarifying remarks are in order. First, natural selection works because only some individuals in a population can survive and this survival is primarily determined by the environmental conditions the population experiences. If environmental conditions change, the types of individuals that are best adapted also changes and, after a time, a population, as a whole, will look different. This is biological evolution at the simplest level: *a change in the characteristics of a population over time.* A very crucial understanding is that no *single* individual ever evolves (this is a simple but surprisingly common misrepresentation of evolutionary thought). Rather, it is the population (a unit) that evolves as *the types of individuals within it* change over time. In evolution, it is the makeup of the population as a whole that is modified.

Natural selection is a biological phenomenon that produces an ordered result. The revolutionary impact of Darwin's ideas was that he offered this naturalistic mechanism that could produce "design," that is, organisms with traits well-suited to their surroundings. An underlying reality in nature is that many individuals are *not* keenly adapted to environmental conditions. Darkly pigmented lizards are poorly adapted and don't survive well at locations with lightly colored rocks because they stand out and are more likely to be taken by predators. Very pale individuals are the best-suited lizards in these conditions. The situation would be reversed in habitats with a dark background. It is interesting that William Paley's famous watchmaker analogy in the late 1700s (an intricately constructed watch implies a purposeful watchmaker) was actually based on a misunderstanding of the real nature of adaptation in the wild. There is always a dynamic actuality that, at any given locality, some individuals possess good traits and others are doomed with poorer attributes. Organisms are not adapted like finely tuned watches; they are more like a smorgasbord of different options.

What was unknown in Darwin's time were the determinants of variation among individuals and the mechanisms of inheritance from one generation to the next. It wasn't until the late 1800s and early 1900s that knowledge of genetics was brought together with the idea of natural selection to form the "new synthesis." We now know that different combinations of DNA instructions cause individual variability. We see that various versions of genes are passed and mixed as parents produce their offspring. So natural selection is really acting to preserve the best-adapted forms and arrangements of genes.

Selection among available DNA combinations can cause striking differences. Darwin was able to document some of this potential by his observations of "selective breeding" or "artificial selection." In these cases, humans, rather than environmental pressures, apply intense selection by determining which individuals have the "correct" traits and are allowed to reproduce to form the next generation. As a wonderfully revealing illustration, humans selecting for specific genetic traits have derived *all* dog breeds, from Great Danes and St. Bernards to Dachshunds and Chihuahuas, all from ancestral wolf stock. It is a little mind-boggling at first glance that a Chihuahua could be the result of successively selecting smaller and smaller individuals as parents for the next generation, but we know that this is how it happened. In the same way, plant breeders have developed an incredible array of vegetables by selective breeding. From a *single species* of wild mustard, farmers have given us broccoli, cabbage, cauliflower, kale, and Brussels sprouts! Many of the radical characteristics chosen by dog or plant breeders certainly would not provide survival advantages in the wild (it is difficult to envision a natural habitat for Dachshunds), but the parallel of this selective process with natural selection demonstrates that there is great potential for population transformation.

The Importance of Mutations

As spontaneous creationists are quick to point out, shuffling of dog genes may create vastly different dog breeds, but they are still dogs. Mutations are alterations in the DNA. These can be a change in a single DNA nucleotide, additions or deletions of many nucleotides or entire genes, or large-scale alterations of entire chromosomes. As far as biologists currently understand, mutational events occur by random mistakes. A large number of mutations are neutral and cause no effect on the physical form or metabolism of an organism. Of course, random mutations within existing functional genes should often be harmful. However, mutations that produce novel favorable results are believed to be the ultimate source for evolutionary change beyond the limits of existing gene options.

The notion that there is continual appearance and accumulation of beneficial mutations is a critical component of modern evolutionary thought. Biologists suggest that favorable mutations, and subsequent new combinations of genes, have arisen and spread through populations, while the constant molding action of natural selection has maximized the quality of surviving individuals. Projected over *very large* periods of time, it is proposed that these general mechanisms have been major players in evolutionary changes leading to the diversity of life on earth. Recent studies in molecular developmental biology (evo-devo biology) demonstrate that there can actually be very substantial evolutionary modifications by reshuffling only *existing* genetic variability, particularly major effects from different combinations of developmental control genes. With that said, the addition of entirely *new* options through beneficial mutations is understood to be a fundamental mechanism for major transformations.

Evolution Completely by Chance

A frequently communicated objection to evolution is that the changes described by scientists could not possibly occur "by chance." It is an absolutely important question to ask whether accumulated mutations and reorganization of gene combinations are an adequate mechanism for what evolutionists claim. However, for there to be effective dialogue on this point, individuals must acknowledge that one part of the evolutionary process, natural selection, is a molding mechanism that absolutely is *not* chance and will always maintain population order and adaptations. The idea that biologists think all the fine-tuned aspects of life have come about *completely by chance* is absurd. I cannot tell you how many times fellow Christians have dismissed me for supposedly holding a position like this.

The most famous version of this argument is a "monkey at the typewriter" caricature. This imagery compares evolution to a monkey randomly striking keys on a typewriter and coming up with perfectly structured sentences or even the works of Shakespeare. But the proposals of evolutionary biologists are absolutely nothing like this. Poorly constructed analogies are easy to attack but they should not sway individuals who have studied the issues more deeply. The crucial point to take forward is that each new evolutionary modification is added to already existing, functional biological organisms that are maintained by natural selection. Each modification or organism is not the result of long sequences of *independent* chance events that would be ridiculously improbable.

The sophisticated arguments of some intelligent design authors also attack the element of chance. ID enthusiasts acknowledge the power of natural selection and mutation but say the odds are too great for these mechanisms to have produced so much complexity or specified information. Although these arguments are far more reasonable in their mathematical approach than the typing monkey analogy, analyses have shown how these reports misrepresent what evolutionary biologist propose and misapply concepts of probability and statistics.[2]

A Good Place for Christians to Stop and Ponder

Christians generally have both theological and emotional difficulties with descriptions of natural selection and genetic mutations. This is entirely understandable in many respects. For example, the discriminating mechanics of natural selection and seeming randomness of mutations appear to exclude God from the evolutionary process. Even more alarming, these mechanisms fundamentally depend on the widespread death of less adapted individuals. I fully expect my Christian brothers and sisters to raise these concerns and have them addressed. However, I also urge believers to not use such concerns to reject what are clearly demonstrated biological mechanisms.

It is important to recall that descriptions of natural processes do not negate the reality of an active and sovereign God. Christians understand that God is continually sustaining the natural order. In addition, we believe that God determines many activities that may appear as chance or unpredictable events to our human perceptions.[3] Believers acknowledge that we do not always know how, when and where God has touched our lives. The important point is that apparent randomness does not mean that a process was undirected. An evolutionary creation view accepts that God used and directed the processes of natural selection and mutations to accomplish his purposes.

Convincing descriptions of natural selection suggest that many Christians must alter strongly held assumptions about what God would or would not do in his creation. Many believers object that a loving God would not allow a process that involves so much suffering and death. It is certainly a significant theological question to ask why God would proceed in this way (like asking why there is so much pain and suffering in the world today). Nonetheless, the Bible itself describes the original creation (Genesis 1-2) as a world not unlike today, with carnivorous animals that kill other creatures, humans using animals and plants for food (i.e., organism death) and the presence of pain (even in the Garden of Eden). In addition, our observations and general logic demonstrate that natural selection is a prominent

2. Shallit and Elsberry, "Playing Games With Probability"; Oloffson, "Troubled Alliance."

3. McGrath, *Surprised By Meaning*, especially 74-81.

mechanism at work in nature. All of this means there is good reason to expect that death has operated since the beginning of life on earth.[4] Christians should be careful to not let emotional responses discount this conclusion.

Perhaps the greatest discomfort for Christians is that natural selection seems so counter to the gospel message. Jesus teaches, "The last will be first, and the first will be last" (Matt 20:16), "Blessed are the meek, for they will inherit the earth" (Matt 5:5), and "Whoever wants to become great among you must be your servant" (Matt 20:26). In contrast, biologists are saying that it is the less adapted, meek and weak individuals that die in the natural world and that the favored creatures are those that dominate others (at least in a biological way). I believe the problem is that Christians are overextending what humans are called to in our relationships with one another and, most importantly, in our right relationship to God. It does not follow that this unique spiritual calling, for those created in God's image, should be expected in all the relationships among organisms in the natural world.[5] In fact, Christians are instructed, with the indwelling of the Holy Spirit, to counter our natural tendencies that often mirror the negative behaviors of other creatures. I invite Christians to embrace our unique place within God's creation and, at the same time, allow the natural world to reveal it's own truths.

Do Favorable DNA Changes Actually Occur?

It is a very legitimate question to ask whether scientists have observed beneficial genetic alterations in nature. We need to begin by considering what is actually being requested here. First, biologists have not been cognizant of this question for very long, a blink of an eye in evolutionary time. Second, many favorable genetic mutations should be expected to cause improvements in the internal physiology or anatomy of particular organisms without being immediately apparent to human observers. Third, advances in molecular biology have only recently allowed us to extensively examine DNA nucleotide sequences to identify new alterations and correlate them with specific organism features.

Even so, biologists have described an array of beneficial adaptations that have arisen in populations and they have, in some cases, linked them to specific genes. A classic example is the appearance of antibiotic resistance within populations of bacteria that did not start with any resistant individuals. The only explanation for this observation is new genetic

4. Murray, *Red in Tooth and Claw*; Schloss, "Evolution, Creation and the Sting," Part 1.

5. Schloss, "Evolution, Creation and the Sting," Part 2.

instructions arising in one or more bacteria. Antibiotic resistance in the bacterium that causes tuberculosis has been shown to be from a single nucleotide change in an identifiable gene.[6] In addition, novel genetic alterations have led to bacteria that can degrade totally new and unnatural, human-produced materials and chemicals such as nylon and toxic PCBs (polychlorinated biphenyl).[7]

Beyond bacteria, biologists have demonstrated that insecticide resistance appeared in a population of blowflies that did not have resistance in earlier generations.[8] A single amino acid change in an existing enzyme allows these insects to degrade the pesticide. In studies of HIV infection in humans, there have always been a few individuals unaffected by the virus, even after multiple exposures. A good deal of research has gone into the mechanisms and specific gene(s) responsible for this condition.[9] This is obviously a favorable genetic alteration in any human populations with a high incidence of HIV. Also, a remarkable beneficial mutant gene that prevents plaque build-up within arteries has been identified within individuals in a small region of Italy.[10] The family members who have this gene do not develop the fatty deposits that commonly lead to heart disease and strokes in other humans. These studies and others like them demonstrate that favorable genetic mutations are a real phenomenon.

Admittedly, these small mutational examples uncovered in the short span of modern humans do not demonstrate all that we need to know to understand large-scale evolutionary changes. This area of study is where a lot of the most invigorating proposals and research are occurring in evolutionary biology. Exceptional discoveries in molecular biology have opened major new understandings of evolutionary novelties in addition to the types of beneficial mutations I've described here (see the later discussion on "macroevolution"). Nonetheless, even as we remain open to new, exciting possibilities in evolutionary processes, the mutation/natural selection model offered by biologists has always contained reasonable natural processes for the development and maintenance of adaptation in organisms. But is there any evidence for the power of evolutionary mechanisms in natural populations?

---

6. Ramaswamy et al., "Single Nucleotide."
7. Prijambada et al., "Nylon Oligomer Degradation"; Copley, "Metabolic Pathway."
8. Newcomb et al., "Single Amino Acid."
9. Eastbook, "Long-Term Non-Progression"; Nolan et al, "HIV."
10. Cedars-Sinai Medical Center, "Defective, But Beneficial."

Population Changes and New Species

Especially important information comes from what biologists call "ring species." The Pacific Coast salamander (*Ensatina eschscholtzi*) is found from British Columbia to northern Baja California. In California, these animals occur along coastal areas and on the western slope of the Sierra Nevada forming a "ring" around the dry Central Valley in the middle of the state (Fig. 15.1). There is a variety of different subspecies (or races) over this range in which the salamanders display noticeable differences in external appearance (i.e., the genetic makeup of individuals within populations differ from one another).

Individuals from *adjacent* populations of *Ensatina* in Oregon and down both the Pacific coast and Sierra Nevada will mate with one another and produce successful offspring. However, something rather remarkable occurs in populations near the southern end of the species' range. At locations in San Diego County, two *Ensatina* populations coexist, one connected to the coastal forms and another most similar to the Sierra Nevada forms. The fascinating thing is that individuals from these two races look very different, ignore one another, and do *not* interbreed.

An evolutionary explanation is that the separated *Ensatina* populations evolved independently in the coastal environments of California and along the western slope of the Sierra Nevada. Sometime in the past, the animals in the two areas extended their ranges south until they now overlap in southern California localities. It appears that animals from the two source regions do not respond to one another because they have diverged so much from ancestral stock through the molding actions of natural selection, and presumably the addition of mutations. This interpretation is bolstered by molecular studies that have demonstrated that coastal and inland populations show more and more genetic divergence from Oregon populations as they are compared further and further to the south.[11]

The *Ensatina* populations form a ring around a physical barrier and there is a lack of interbreeding in one area where the ring connects. These observations indicate that the molding of individual populations to local conditions can lead to significant divergence, even to the point that populations look very different and become reproductively incompatible with animals that share a common ancestry. This is precisely the method by which modern evolutionists suggest that new independent species are most commonly derived. The two southern California populations of *Ensatina* are related through past ancestors, and yet, the individuals in the populations

---

11. Moritz et al., "Evolutionary Relationships"; Wade and Schneider, "Taxonomy."

have evolved to the point of biologically distinct entities! There are many other examples of ring species in nature.[12]

Ring species provide strong evidence that evolutionary mechanisms not only alter the characteristics of populations but also cause the appearance of new species. A helpful parallel is to consider how all dog breeds are genetically related as one species. Individuals from most breeds can freely mate with one another (we could line up dog types by size and similarity in form). However, the most extremely modified forms, Great Danes and

Figure 15.1 — *Ensatina eschscholtzii* **salamanders in Oregon and California, a ring species.** Populations of these salamanders occur along the California coastal mountains and in the foothills of the Sierra Nevada range to form a ring around the dry Central Valley. Individuals from adjacent

12. e.g., Irwin et al., "Speciation in a Ring."

populations freely interact and interbreed over most of this range. However, in specific southern California locations, salamanders from separate coastal and inland ancestries do not interact or interbreed.

Chihuahuas for example, are so different that they are incapable of successful reproduction. What would biologists conclude if all intermediate dog breeds were eliminated? If this were to happen, the genetic future of Great Danes and Chihuahuas would never intermingle and they would represent two distinct species.

The Significance of Hybrids

Additional evidence that new species are derived by the genetic divergence of separate populations comes from the study of hybrids. A "hybrid" is produced when individuals from physically and behaviorally different species interbreed and give rise to an offspring. There are numerous examples in nature. The fact that a hybrid can be formed at all reveals a genetic connection between very distinct parent types. This makes sense if the parent species came about by evolutionary divergence in the past. One of the best-known hybrid examples is a mule, the product of a horse and donkey mating. Since mules are sterile, there is no blending of the parent species.

Wonderful illustrations of hybridization come from western-eastern U.S. bird pairs that have been well-studied where they overlap in the mid-continent. A sample of these is listed in Table 15.1. Some of the species pairs (bluebirds and meadowlarks) have been observed to produce hybrid offspring only on rare occasions. Other pairs (buntings and orioles) frequently interbreed, but only in narrow hybrid zones that extend from the Canadian prairies through Colorado, Nebraska, and Kansas. Finally, some of the matching west-east bird types (flickers and warblers) have been found to freely interbreed wherever they meet.

It has long been thought that the bird pairs mentioned here arose during the Pleistocene ice ages (dating back to 2 million years ago) when a glacier down the center of the continent isolated western and eastern populations of many species. Molecular studies have supported this time frame for some species[13] and have suggested an even earlier separation for other species (up to 5 million years ago).[14] The existence of hybrids in these examples supports the notion that the bird pairs came from common ancestral stock and that the separated populations diverged in characteristics

13. Weir and Schluter, "Ice Sheets."
14. Klicka and Zink, "Importance of Recent Ice Ages."

while isolated in western and eastern parts of North American. Biologists conclude that the divergence happened through rearrangement of genetic combinations and accumulated genetic changes under differing environmental conditions.

The hybrid examples in Table 15.1 demonstrate that species pairs diverged to different degrees while they were separated. The bluebird and meadowlark populations became so different that hybrids are rare and infertile. This means that the parent species will remain distinct. On the other hand, the separated flicker and warbler populations currently produce such viable and fertile hybrids that the "species" are slowly blending back together. This indicates that genetic changes when the populations were apart were not extensive enough to fully divide the stock into two entirely distinct entities. The bunting and oriole species pairs show an intermediate condition with a limited number of hybrids with reduced vigor and reproductive success only in the middle of the continent.

Table 15.1 — Selected species pairs with range overlap and the presence of hybrids in the central United States.

Western Form		Eastern Form
Western Bluebird	< hybrids rare, infertile >	Eastern Bluebird
Western Meadowlark	< hybrids rare, infertile >	Eastern Meadowlark
Lazuli Bunting	< hybrids limited, infertile >	Indigo Bunting
Bullock's Oriole	< hybrids limited, less fertile >	Baltimore Oriole
Red-shafted Flicker	< freely interbreed >	Yellow-shafted Flicker
Audubon's Warbler	< freely interbreed >	Myrtle Warbler

A broader study of reproduction in nature reveals that hybridization between distinct species is actually very common. This includes known hybrids among related forms from within just about every imaginable general type of animal (e.g., between different species of shrews, between numerous species of ducks, and likewise for frogs, fishes, flies, and many more). Plants are also known for their frequent hybridizations. In fact, botanists generally expect to observe at least a few hybrid individuals in areas of overlap between two similar species. For example, many of the 45 species within the *Ceanothus* genus mentioned in chapter 13 produce offspring of mixed parentage. However, the critical attribute of hybrid offspring is that they typically possess inferior survival capabilities and/or reduced fertility in most situations. This means that the parent forms remain distinct biological groupings even as the hybrids disclose to us an ancestral evolutionary connection.

Ring species and hybrids are consistent with what we expect under current models of evolutionary change. This gives evolutionary biologists good reason to conclude that natural selection, mutations and other mechanisms have been an important force in modifying organisms and generating new species over time. It is also sensible to extrapolate that these events operated in significant ways over very long periods of earth history. There is no particular reason why they couldn't produce entire lineages, such as all 45 species of *Ceanothus* species in western North America or the 432 species of tyrant flycatchers found throughout the New World. These mechanisms also should have played a fundamental role in major evolutionary transformations like those revealed in the evidences of preceding chapters.

Interpretations

Essentially all spontaneous creationists acknowledge that natural selection can operate to change population characteristics, at least to some extent. Most also accept that it can mold separate populations of one species into reproductively separated species (i.e., speciation). However, the majority suggest there are limits to this "microevolution." There are some intriguing implications that accompany these positions.

First, an acceptance of significant microevolution necessarily implies a considerable period of time for such evolutionary changes to take place. Although evolutionary biologists believe there can be occasionally rapid evolution, none of them would propose that lineages such as the Darwin's finches or the *Ceanothus* species of the western U.S. could be derived in the 10,000-year time frame most often associated with young-earth creation. As one reference point, the eastern and western U.S. bird species pairs just mentioned are estimated to have diverged two to five *million* years ago. Many spontaneous creationists understand this dilemma and grant that the earth is very old. Most old-earth creationists and intelligent design supporters concede there has been substantial biological evolution and agree with the several-billion-year time frame understood by modern scientists. These positions can become increasingly similar to evolutionary creation.

However, the most common young-earth creation characterization is a logically inconsistent acknowledgement of microevolution events within "kinds" along with an insistence that they occurred within a 10,000–12,000-year period or an even shorter timeframe since the global flood of Noah. I mentioned this odd situation in chapter 13. This isn't biological evolution and it is not really based on any known mechanism. There is no evidence in natural world or in the Bible that miraculous transformations advanced this quickly *while humans were present on the earth*. In comparison, pale-ontologists report from the fossil record that there was an *unusually rapid* production of antelope species limited to Africa over a 400,000-year period between 2.9 and 2.5 million years ago. This coincided with oscillations in climate that continually altered savannah and grassland habitats.[15]

A second implication that follows if spontaneous creationists accommodate very much microevolution is that it actually validates a large portion of the work of modern evolutionary biologists and, ironically, the majority of Darwin's presentations in *The Origin of Species*. If it is acknowledged that the Galápagos finches, the U.S. bird species pairings, or the diverse *Ceanothus* forms across the West each evolved from an ancestral stock, then an individual is accepting *quite a lot of evolution*! This amount of modification in organisms is certainly very different than anything imagined by Christians in the past who simply assumed species were distinct fixed entities.

One has to wonder, if it is acceptable to acknowledge significant microevolution within broad biblical kinds, what really are the biological boundaries of each kind? Our general observations lead us to understand general categories such as "ducks" or "oaks," but closer studies of living species and forms in the fossil record reveal that our human-constructed biological categories actually blend into one another. For example, bird watchers know that living species of warblers, sparrows, and blackbirds show a continuum among the types so that these are not really obvious designations. Also, it is difficult to delineate a clear distinction in fossils between fish and amphibians or between feathered, flightless dinosaurs and birds.

Finally, the fact that many spontaneous creationists acknowledge substantial microevolution is, to me, both encouraging and inviting. It demonstrates that biblical views previously held by some individuals about the age of the earth and the meaning of "kind" can be revisited based on evidence revealed in God's natural revelation. It means that our interpretation of the Bible can be altered and improved by this evidence. Is it possible that we can move further toward a climate in which the wonderful stories of biology can be more fully merged with biblical truths? This is, of course, the major goal

15. Purves et al., *Life*, 492.

of this book. I remain convinced that in evolutionary creation a deep and comprehensive understanding of biology and a full and uncompromising biblical faith can be integrated into a single meaningful worldview.

Macroevolution

Historically, many biologists have suggested that we can extrapolate the molding actions of mutations and natural selection to explain the grand scale of change at the "macroevolution" level. In this view, there really is no precise distinction between micro- and macroevolution other than a vague notion about the amount of time involved and the degree of evolutionary transformation. In fact, biologists often avoid using "micro" and "macro" designations. In contrast, spontaneous creationists have charged that showing that mutation/natural selection mechanisms can operate in the short term does not imply that they are wholly adequate over the long term. I believe biologists have been mostly justified in defending the power of mutations and natural selection, but, at the same time, it has never been certain that we understood all the natural mechanisms needed to explain the grand story of life. Spontaneous creationists have rightly criticized scientists who have implied that our knowledge base was wholly adequate. A much better approach is to continue to express tentativeness and acknowledge that there is much more to be discovered about evolutionary processes.

The appropriateness of this attitude has been pointedly illustrated by innovative techniques in molecular biology that have opened up brand-new ways to examine and understand evolutionary history. In recent years, scientists have precisely mapped entire DNA genomes and unraveled incredible complexities of cellular mechanics. There have been numerous discoveries that have both surprised and delighted scientists.

One easy-to-understand revelation is that duplicated genes occur much more frequently along chromosomes than was ever expected. This has very significant consequences. In the past, we typically viewed mutations primarily as disruptions in currently functioning systems. But a mutation in a duplicated gene can occur without harming an organism because the original gene still functions and gives rise to the normal protein product. As discussed in chapter 14, an altered duplicated gene may become a disabled pseudogene and accumulate additional errors over time. However, a mutation in a duplicated gene could alternatively lead to a new protein product that performs a valuable function *in addition* to what is accomplished by the original gene. The consequence is that random mutations, including potentially favorable ones, can occur relatively frequently without interrupting working cell mechanisms. This is very different than previous notions that a beneficial DNA change would have to occur within a single version of a functional gene. The existence of duplicated genes greatly increases the likelihood of favorable alterations in the evolutionary past.

Researchers are investigating other molecular mechanisms of evolutionary change that are beyond the scope of this book. These include (1) jumping elements of DNA (transposons) that can alter the regulation of protein-coding genes, (2) conversion of nonfunctional DNA into operational, beneficial genes, (3) the origin of new genes from messenger RNA sequences, and (4) co-opted genes from parasitic viral DNA.[16]

Some of the most important discoveries have been in genetic mechanisms that control embryonic development. Recall that the form and operation of adult animals are actually determined earlier during the unfolding of embryo instructions. Molecular biologists have uncovered a surprising commonality among animals with the revelation that a relatively small number of genes direct crucial embryonic events. These "developmental control genes" turn on in sequence so that there is a cascade of instructions to regulate the expression of other genes. This determines things such as the front and back end of an animal, partitioning of the embryo into segments (like the bony vertebrae down the back of a dog or the sections in a centipede), and controlling what body parts develop in each segment. Differences in where these control genes are expressed within the embryo, the timing of their expression, and the amount of product from each gene have been shown to play major roles in the resulting adult form.

Let me give an example to illustrate the point. In snakes, a particular gene, *Hoxc8*, is expressed in the "shoulder" part of the embryonic form where forelimbs develop in lizards. The *Hoxc8* gene product *prevents* the formation of limb buds in snakes and, of course, snakes have no front legs. The remarkable thing is that the same *Hoxc8* gene is present in the shoulder region of lizard embryos. The difference is that the *Hoxc8* gene is *not* expressed in lizards and this is what leads to the growth of limb buds. In other words, it is the *inactivity* of *Hoxc8* that allows for forelimbs in lizards! Within the developing lizard limb buds, subsequent control genes turn on or off, in sequence, to direct the detailed form of the legs. These observations mean that the *Hoxc8* gene, a single DNA segment, is determining the complete presence or absence of forelimbs!

Another illustration for the significant influence of control genes is the extremely long neck of giraffes. Giraffes have seven neck vertebrae, just

16. Fedoroff, "Transposable Elements"; Kaesmann, "Origins"; Venema, "Understanding Evolution," Parts 2 and 3.

like almost all mammals, including humans. The way the animal's neck becomes so elongated is that particular developmental genes stay turned on for extended periods of time compared to other mammals. It is not that numerous unique genes are required to form the unusual giraffe anatomy. Instead, it is the result of relatively few differences in the operation of the same control genes shared among all mammals.

Can you see how this new understanding alters our notion of beneficial mutations? Just think if shape change, reduction, or even elimination of animal features requires only a relatively small number of alterations at developmental control genes rather than a large series of accumulated genetic alterations. Molecular biologists and evolutionary biologists have recognized that this is fertile territory and the field of evolutionary developmental biology (evo-devo biology) has blossomed.[17] The take-home message is that genetic changes that alter the regulation of developmental steps are the likely source of many major evolutionary changes. This means that organism differences may not be so much the presence or absence of particular genes but rather the regulation of where, when, and how existing developmental genes are expressed. It is highly significant that these processes contrast with the typically perceived evolutionary progression of small, accumulated steps.

There have also been searches for other novel natural mechanisms or conceptualizations of evolutionary change. There is increasing evidence from the study of microorganisms that major transfers of DNA segments among entirely unrelated organisms probably played an important role, especially in the distant past (e.g., during the Precambrian and Cambrian periods). We now know that bacteria will pick up stray DNA from the environment, incorporate it into their own genome, and pass it into future generations of progeny cells. Some scientists are investigating the transmission of characteristics across generations based on control molecules associated along side DNA molecules (rather than on differences in the DNA itself). Epigenetic inheritance, as it is called, does not appear to last over multiple generations but it is still possible that it may have had a more important role in evolution than is currently understood.[18]

No less staunch a defender of evolution than Stephen J. Gould argued in his monumental work, *The Structure of Evolutionary Theory*, that there is too much extrapolation of what we know about small-scale genetic changes

17. Arthur, "Emerging"; Carroll, *Endless Forms*; Sommer, "Future of Evo-Devo"; Muller and Newman, "Special Issue."

18. Chandler, "Paramutation"; Jablonka and Raz, "Transgenerational"; Rando and Verstrepen, "Timescales."

and selection processes to explain all of macroevolution.[19] He suggested that biologists should focus more on selection processes at the level of entire groups, populations, or species. He also noted that biologists have probably erred in assuming that natural selection is the nearly exclusive mechanism of differential survival instead of studying more random processes like genetic drift and adaptive sorting that might have significant evolutionary effects.

Gould has been followed by a number of scientists that continue to probe for a more comprehensive understanding of natural mechanisms in evolutionary theory. New proposals for an "extended evolutionary synthesis" are being hotly debated within the scientific community and among philosophers of science[20] (as a side note; they are also erroneously utilized by spontaneous creationists to claim that evolutionary theory is in turmoil). It is probably helpful for Christians to know that well-respected biologists who are also Christian evolutionary creationists are part of this ongoing discussion.[21]

In conclusion, spontaneous creationists are incorrect when they imply that biologists have no reasonable scientific explanations for macroevolutionary change. All living things are based on the same DNA with the same rules for activating and expressing the instructions contained on those molecules. It is not difficult to see that by various chromosomal additions or deletions, and changes in DNA bases, it is possible to get from the specific instructions of one organism to any other organism. There is no inherent limitation indicating fixed "kinds" and no obvious distinction between changes accomplished at the microevolution level and those required for macroevolution. I invite all readers to conclude that the natural world not only tells a remarkable story of an evolutionary past but also continues to reveal fabulous details about how evolutionary progressions have occurred.

19. Gould, *Structure*, 566–91.

20. Koonin, "*Origin* at 150"; Pigliucci and Muller, *Extended Synthesis*.

21. Conway-Morris, *Life's Solution*; Conway-Morris, "Darwin"; Murphy and Schloss, "Biology and Religion"; Schloss, "Sting of Death," Part 3; Murray and Schloss, "Evolution, Design."

PART V

Reading The Bible
With Evolution In Mind

16

Reconciling Scripture and Evolutionary Theory

AFTER ALL THE BIOLOGICAL details in the last several chapters, it is time to focus on the topic of particular concern for most Christians: How can evolutionary theory be accepted when there are contradictions with the Bible, especially with the initial accounts in the book of Genesis? The conflict between a long evolutionary history and six days of creation is strikingly obvious. And what about the stories of Adam and Eve, the Garden of Eden, and the biblical flood? It is because of these issues that many Christians feel that they must reject at least some, if not the entire, evolutionary story described by scientists. These concerns are at the very center of the creation versus evolution controversy.

My heart and mind have always told me we are missing something terribly important in these presumed conflicts, and many other Christians join me in this awareness. To begin with, we should be free to embrace the long church tradition that both the words of the Bible and evidences in nature are valid revelations from God to us. If God is the author of both, we can trust that there will be no real disagreement. Put a different way, we cannot comfortably argue that God created one reliable source of information in the Bible and created a second conflicting, unreliable source in nature. This means we should use Scripture to rightly interpret what we observe and experience in nature and, at the same time, allow the natural world to inform us how to rightly interpret the Bible. This statement may seem foreign to evangelicals seeped in an undue emphasis on the Bible as a singular authority, but it is really both a fully reasonable and orthodox Christian approach. In fact, we may proceed with complete confidence that the biblical text and scientific information will beautifully complement one another in informing us about God and his ways.

I have explained the persuasive scientific position for a very old earth and long evolutionary ancestries. Some Christians still hope to disregard these conclusions by suggesting our scientific understandings will change, but this hugely underestimates the quantity of diverse evidence and fails to comprehend how valuable evolutionary theory is within biology. Evolutionary theory is an encompassing scientific framework that is here to stay. If skeptical Christians acknowledge even some of the evidence, the proper response has to be to allow oneself to be challenged regarding what, if anything, the Bible is actually saying about such things. I have communicated that the natural world presents a unifying message. As odd as it may seem to many conservative Christians, this means that it is ultimately up to believers to discover how God's Word is in harmony with evolutionary understandings.

By making this suggestion I proclaim loudly and clearly that I am not relegating the Bible to a subordinate position. Scripture is the primary means by which a Christian's worldview is defined, and its truths guide us in our daily lives. I am well aware that believers must be careful not to alter what the text says to make the Bible mean something we think is easier to swallow or to simply rationalize a preconceived position. We should expect Scripture to challenge us and we must, as a principal rule, wrestle directly with what it says. At the same time, it is very important to acknowledge that individuals who are united by their faith in biblical inerrancy often differ in their interpretations of what biblical authors intended.

Proper Bible interpretation (exegesis) must take into account all possible sources of information to gain clarity on what God is actually communicating through his Word. We will always use internal biblical consistency as a guiding principle. But we also need to consider other features, including the author's focus, the historical setting and cultural context in which the words were written, and the literary style of the text. There is a large body of scholarly work devoted to these features. We also live in an unusual time of new scientific information that has become absolutely relevant to discerning what the Bible is communicating about creation events. In the discussions that follow, I proceed with the conviction that the Bible is God's specific revelation to humanity that must be understood and not easily dismissed. At the same time, I embrace the natural truths revealed by my colleagues in the biological and physical sciences.

Learning from the Past

Evaluating biblical meaning in light of scientific information is neither sacrilegious nor damaging; it is, instead, the proper path to strength and relevance for the Christian church. We would be wise to learn from the infamous astronomical debates of the 16th and 17th century. Increasingly detailed observations of moving celestial objects led investigators of the time to suggest a sun-centered solar system. In contrast, there was good biblical reasoning that dictated that the earth was at the center of the universe and that the sun and other heavenly bodies moved around it. For example, Genesis 1 seemingly communicates a focal location for our planet as it states that the sun, moon, and stars were placed in the sky above the earth. Also, a number of Bible verses clearly describe the sun *moving* relative to the earth. Psalm 19:6 reads, "It [the sun] rises at one end of the heavens and makes its circuit to the other." On top of this, it appeared self-evident (just as it does today) that celestial bodies move through the sky around a central, stationary earth. When the observations and calculations of Copernicus, Kepler, and Galileo suggested that the earth revolves around the sun at the center of our solar system, some members of the Catholic Church were not ready to abandon their established position. Of course, all individuals eventually conceded that the inerrant biblical text had been *misinterpreted* with regard to what it communicated about these scientific matters.

It should undoubtedly be noted this does not demonstrate that Scripture is inaccurate or wrong. We now conclude that verses like Psalm 19:6 address the movement of the sun from an earthbound perspective that would be easily understood by readers at the time of their writing, as well as by us today. We understand with our modern knowledge that these verses were never intended to convey a scientific reality that the sun actually moves around the earth. The mistake occurred when some church intellectuals defended a scientific position based on biblical interpretations and personal perceptions and tried to dictate to astronomers what they could discover as valid information. Even further, they held on to their position despite mounting scientific evidence to the contrary.

This legacy is still powerful ammunition for those who choose to reject Christianity based on a perceived inability of the church to be relevant in a modern world. But Christians have always understood that, although the Bible is God's revelation to reach the hearts and minds of people over all the ages, each section was written in the distant past to the knowledge base of ancient cultures. As such, the Bible has never been a scientific document. It was especially not written to address questions raised by modern investigations. Copernican understandings compelled people to reassess their biblical readings and accept that honest scientific discoveries can seem quite foreign and challenging to our own personal perceptions. The lesson is that the Christian church remains engaged only as it enthusiastically acknowledges valid scientific findings and embraces a general truth that the Bible's primary purpose is in its theological message, not in a scientific one.

The resistance to a sun-centered solar system is strikingly similar to how spontaneous creationists are responding to evolutionary theory today. First, opposing arguments are largely motivated by biblical readings that dictate what should be discovered and what is allowable in science. Claims that the earth is young, that evolutionary transformations cannot explain life's diversity, or that God should be detectable in the complexities of nature, are all based on personally chosen Bible interpretations. Second, there is an appeal to our short-term experiences that seem to show that organisms do not evolve and that remarkable features in nature couldn't have arisen by natural mechanisms. But the supportive evidence behind evolutionary theory has been mounting for years.

In parallel with altered thinking after the Copernican revolution, I suggest that modern evolutionary findings should compel Christians to reevaluate conflicting readings of Scripture. Part of the difficulty in this transformation is a frequent misapplication of the principle of "inerrancy." In this error, believers assume their current Bible interpretations are accurate and inerrant and, therefore, they should not be changed. In contrast, we should properly understand that *the Bible is inerrant* in what God intended it to communicate; that is, God does not err and he is the essence of truth. It is an entirely different thing to claim that any one particular *human interpretation* of a Bible passage is inerrant. What we really know is that God's full intentions are not always evident. We may follow the path of notable Christian leaders through the history of the church who considered a search for the deepest meanings within God's Word an ongoing process.

A particularly instructive aspect of the Copernican/Galileo story is that the Christian church has more than survived the acceptance of a heliocentric solar system. We no longer view the universe the way we did in the 16th century, but this is no blow to our perceptions of God or the Christian gospel. In fact, scientific discoveries reveal a God that is even more amazing then previously perceived. We now know that our solar system is a small part of a universe with over 200 billion galaxies, all of which Christians assert are the handiwork of God! It is important to remember that although God is unchanging, our understanding of him does not have to stay unchanged. The church is strongest when its members are open to a dynamic search to better know the Holy God of the universe who will never be wholly knowable.

Making Changes

A good number of evangelical Christian scholars, pastors and teachers have come to appreciate the challenge and opportunity I am expressing in this discussion. They are engaging with modern science and applying a biblical worldview to put evolutionary information into an encompassing Christian context. By these actions, Christian laypeople are being equipped to more effectively embrace and share their faith in our science-oriented society. I eagerly invite others to join the effort.

Even with these positive signs, there are various obstacles to progress within many evangelical churches. One factor is that pastors and other church leaders typically give scientific information a low priority within their busy and complicated lives. This is understandable in many respects. These individuals face enormous demands in their calling to shepherd fellow believers. In addition, most have limited science education and no particular affinity for scientific studies. This means that challenges from evolutionary biology put most pastors and teachers in a difficult situation. I really sympathize with their predicament. Before I go on, I want to fully express that these leaders deserve great thanks and they should be held "in the highest regard in love because of their work" (1 Thes 5:13).

At the same time, I have sat through numerous sermons during which well-meaning teachers presented simple interpretations of Genesis that strongly conflict with modern science (as well as with current contributions in biblical studies). These include references to six literal days of creation, a perfect Garden of Eden without pain or death, and descriptions of a world-wide flood. I have also been aware that many listeners (especially those with scientific backgrounds) did not agree and were uncomfortable that there was not at least some mention of alternative approaches to these Scripture passages. In the majority of instances, this disconnect between the reflections of the pastor and the perceptions of some listeners was left unacknowledged and unresolved as if it has no consequences. But this means it is up to laypeople to individually sort out the conflict between their church culture and scientific knowledge. An incredibly important observation is that, in most cases, including sensitivity to different natural interpretations would not have altered the theological messages in these sermons! I suggest that many pastors can more directly model to their congregations that there are different legitimate Genesis interpretations among devoted biblical Christians.

Even further, interpreting Bible creation events stands out in a very distinct way compared to bringing forth meaning from most other parts of Scripture. This is because creation accounts immediately draw us to comparisons with scientific knowledge from God's natural revelation. This means that pastors and leaders have at least some unique obligation to understand the science of our day, even though they are primarily trained to rely upon biblical studies. We have seen that notable Christian theologians have taught us to actively engage with academic pursuits and wrestle with discoveries. Tim Keller, influential pastor and author from the conservative Presbyterian Church in America, puts it this way:

> If I as a pastor want to help both believers and inquirers to relate science and faith coherently, I must read the works of scientists, exegetes, philosophers, and theologians and then interpret them for my people. Someone might counter that this is too great a burden to put on pastors, that instead they should simply refer their laypeople to the works of scholars. But if pastors are not "up to the job" of distilling and understanding the writings of scholars in various disciplines, how will our laypeople do it?[1]

I realize that many evangelical pastors and leaders will protest that they *have* investigated scientific claims for biological evolution, and have rejected them. However, a common reality is that many have primarily or exclusively become familiar with the arguments of young-earth creation and/or intelligent design. Since spontaneous creationists represent a very small percentage of scientists, this means evangelical leaders have often received only isolated minority views. As I have noted, the mainstream scientific community has refuted spontaneous creationist arguments in great detail and consider them soundly rejected. Even so, major references in systematic theology, influential teaching at Christian seminaries, and important sharing among church pastors repeat these arguments as if they are fully valid.

I earnestly suggest that Christian leaders take more notice of the very large number of educated, devout Christian biologists who thoroughly understand and agree with evolutionary conclusions. The latter can fruitfully explain how the arguments of young-earth creation and ID have little merit within science and are so frequently detrimental to sharing the gospel of Jesus. In sum, I urge pastors and teachers to gain a solid knowledge of why so many scientists embrace evolutionary theory. In turn, I encourage them to rally believers to engage with evolutionary science and a society that largely, and legitimately, does not dispute it. I make this suggestion not because I idolize scientific accuracy, but for the health and effectiveness of the church. Christians will always praise God for making the entire natural world, and scientific discoveries will never alter this proclamation. This is not a case of

---

1. Keller, "Christian Laypeople," 3.

being "blown here and there by every wind of teaching" (Eph 4:14) but rather an opportunity to strengthen our faith. I suggest that evolutionary creation allows Christians to more fully appreciate God and all he has done. Even further, it best prepares believers to offer the transforming message of Jesus.

A Universal Doctrine of Creation

Fortunately, regardless of how Christians read the first chapters of Scripture, there are certain elements in the creation accounts that theologians have commonly recognized as key communications from God to his people. These comprise a core doctrine of creation around which all believers can express unity. By "doctrine" I mean understandings that have been the steady expression of Christians throughout the history of the church. Before I proceed to an analysis of the Genesis text in light of evolutionary theory, I believe there is great value in celebrating aspects of this creation doctrine. These include the following:

There is one God. There are not many gods or spiritual beings of importance; rather, there is one true God over all things. God made all things in heaven and earth, both visible and invisible. The "Apostle's Creed," a widely accepted foundational statement of the early Christian church, contains the succinct statement, "I believe in God the Father almighty, maker (or creator) of heaven and earth." In addition, God instilled his creation with order. Therefore, the world is not the result of meaningless physical processes; it is the intended plan of an all-powerful God. God exists apart from the physical world and he is not limited by the parameters of the things he created. Instead, he is beyond the constraints of space and time. God is sovereign over all that exists and all things depend on his sustaining power (Col 1:17). Since God is behind our existence, life has purpose, meaning, and an ethical structure as defined by him. Finally, God's creative actions culminated in humans (Adam and Eve) whom he designed to enter into a unique relationship with him and be his representatives on earth. We are created in the image of God so that we, apart from all the rest of creation, might share this distinct and special relationship. The way the *Westminster Confession of Faith* puts it, the chief and highest end of man is "to glorify God, and fully to enjoy him forever."[2]

The preceding statements are significant explanations about the world in which we live! There are certainly additional theological implications in the initial chapters of Genesis, including the pattern of a Sabbath week, the inherent goodness of God's creation, the intimate bond and purpose in

2. *Westminster Confession of Faith*, 153.

marriage relationships, and the role of humans as stewards on earth, but I respectfully leave their full elaboration to others more qualified than I.[3]

A noteworthy aspect of this doctrine of creation is that it communicates important foundations of Christian faith without reference to God's *methods* or *timeframe* of creation. In this sense, believers may see that the Genesis accounts contain much of what is theologically important to us without specifically delineating God's creative acts. No scientific conclusions will ever alter these fundamental truths and Christians may come together on these core elements. This should be so for unity within the church, but also for the sake of non-Christians who should be allowed to believe about such things whatever they are convinced of by their intellect, so that there will be no stumbling block between them and Christian faith.

Approaches to Reconciling the Bible And Evolutionary Theory

The discussion in the upcoming chapters will cover the most controversial aspects of early Genesis. My comments will not be exhaustive. They are intended, instead, to open various avenues of thinking for bringing together biblical meanings with evolutionary understandings. It is my hope that these issues will be fleshed out more fully through other resources and personal investigation.

To begin, it is important to be clear that strict, word-for-word, literal readings of early Genesis are not a viable option. Of course I understand that for some Christians, to question a Genesis reading of special creation over six 24-hour days is plainly heretical. Even so, I have explained how this position does not line up with what the natural world tells us about the earth's history. Even more important, an analysis of the Bible itself does not support such a one-dimensional approach.

First, there are prominent parts of Scripture, such as Proverbs and the Psalms, which are clearly poetic and metaphorical in nature with rich use of allegory, analogy, and other figurative expressions. It is simplistic to claim that we are incapable of distinguishing passages like this from verses of straight narrative prose. Theologians have noted through the history of the church, and long before the advancement of evolutionary science, that *all* of the early Genesis chapters display distinctive literary structures and metaphorical elements. These features indicate that a full understanding of the texts will be revealed when we allow for figurative communications. This does not mean the chapters lack a basis in real historical events, but it

---

3. e.g., Collins, *Genesis 1–4*, 129–32 and 269–78; Alexander, *Creation or Evolution*, 27–34.

does mean they should not be taken as word-for-word factual accounts. C. John Collins, Old Testament professor at Covenant Theological Seminary, provides helpful perspective:

> When the Biblical authors describe people and events for us, they are aiming to give us more than simple facts; they want to capture our imaginations, to enable our whole persons to lean into life with a vigorous faith and a zeal for goodness.[4]

Second, a plainly literal interpretation of Scripture is an unreasonable requirement that is impossible to apply even to the narrative parts of the Bible. For example, within the laws and regulations detailed in the book of Leviticus, verse 11:20 reads, "All flying insects that walk on all fours are to be detestable to you." Insects are characterized by the possession of *six* legs, but a purely literal interpretation would compel us to insist that the insects around us actually have only four! We are far better off accepting that "walk on all fours," meant something different to early Israelites than what is implied by a strict literal reading.

Another example is the use of phrases in both the Old and New Testaments that flow from past perceptions that the earth is flat. These include "in heaven and on earth and under the earth,"[5] plus references to the "ends of the earth."[6] Are we to argue for a literal reality that the earth is flat based on these texts? We must acknowledge that the primary intent of the Bible is to make statements about God and our relationship to him, not literal details about scientific matters.

A key advancement in Old Testament studies over the last century has been appreciating that the early Genesis passages were written in the context of surrounding cultures in the ancient Near East.[7] Many of the communications are distinctive, defining views about the God of Israel and his covenant people in contrast to beliefs in other groups (the doctrine of creation above). Even so, people across all these cultures shared common understandings about the world in which they lived; including a flat earth, an earth fixed in position, and the waters of heaven held back by a dome-like firmament. These beliefs are expressed within the Bible even though we know today that they are scientifically inaccurate.[8]

---

4. Collins, *Adam and Eve*, 20.

5. Phil 1:10, Rev 5:13.

6. Ps 19:4, 48:10; Isa 24:16, 49:6.

7. Enns, *Inspiration*, 23–70; Niehaus, *Ancient Near East*; Oswalt, *Bible Among Myths*; Walton, *Ancient Near East*.

8. e.g., reference to the firmament remains in Genesis 1:7 of the *King James Bible*: "And God made the firmament, and divided the waters which were under the firmament

This leads us to a crucial observation: God did not correct this faulty science when he inspired biblical authors. If we pause for a moment, it is easy to see that modern scientific accounts would have been meaningless to an ancient audience. Even more important, unintelligible modern descriptions would have overshadowed or confused the real theological message and purposes of the passages. Theologians from Augustine to current scholars have highlighted this significant aspect of Scripture. John Calvin, the prominent Reformation leader, noted in his *Commentary on Genesis*:

> Moses wrote in a popular style things which without instruction, all ordinary persons, endued with common sense, are able to understand . . . Had he spoken of things generally unknown, the uneducated might have pleaded in excuse that such subjects were beyond their capacity.[9]

Calvin also famously instructed, "He who would learn astronomy, and other recondite arts, let him go elsewhere."[10]

There are two important lessons here. One, it is wrong to assume that the God of the Bible, a God of truth, should necessarily communicate modern scientific accuracy within pages of Scripture that were written thousands of years ago. This directs us to look for deeper theological meanings that were relevant in the ancient past, as well as today. Two, we must conclude that the initial Genesis chapters were not intended as a straight literal history.

Young-earth creationists often argue that we should read Genesis as historically accurate in every detail or we open the door for loose, figurative analyses where the words of the Bible become manipulated to mean whatever the reader wants. The problem is we have many indications that ancient audiences expected the stories to have artistic frameworks and that they deemed the best accounts were those that communicated the heart of the matter without concern for exact chronologies or scientific precision. Their primary questions were about ultimate meanings and purpose, not about mechanical origins. The efforts of young-earth adherents are well intended, but it is a strange irony that they are attempting to honor and protect the Bible by promoting a type of excessive literalism that the biblical authors would not even understand.

There is a general consensus among Old Testament scholars that Genesis must be read and interpreted through the eyes of a capable reader at the

---

from the waters which were above the firmament: and it was so."

9. Calvin, *Commentary*, Vol 1, 1:6.

10. Ibid., Vol 1, 1:6.

time the accounts were written. John Walton, from conservative Wheaton College, expresses the point well:

> The Old Testament *does* communicate to us and it was written for us, and for all humankind. But it was not written *to* us. It was written to Israel. It is God's revelation of himself to Israel and secondarily through Israel to everyone else. (italics in the original)[11]

This means that good interpretations of Genesis will not only translate the written words, but also grasp the significance of those words within the beliefs and conventions of ancient Israelites. We may remain mindful that these are God-inspired writings that may contain truths hidden from ancient participants that are perceptible to later readers. However, our primary understanding must be that God communicated within an ancient cultural context to teach principles for people of both the past and today.

One thing we should *not* do is claim the first books of Genesis were combined in a haphazard fashion within an unenlightened or naïve culture. For example, it is sometimes suggested that Genesis 1 and Genesis 2 contain two separately written, conflicting creation stories that were simply pasted together. But why would the early Israelites allow an obvious disagreement between the first two chapters of the defining Pentateuch? It is likely that this is a conflict only in the minds of modern readers. It is far more reasonable to conclude that the first chapters of Genesis are skillfully and purposely organized to convey significant, worldview-defining messages about God and his covenant people and that ancient cultures understood these accounts in a coherent and meaningful way.

We are left with two main approaches for bringing together Genesis and evolutionary discoveries. One option is to understand that early Genesis is almost *entirely figurative or symbolic in nature*. In this case, we reason that a biblical author communicated in a metaphorical fashion within the culture of the time. We then look for eternal truths expressed by analogy as is commonly done with poetry or great storytelling. If a biblical passage is understood this way, the expressed truths about the general character of God, his creation, and our relationship to him are recognized to be of great significance and we are invited to relate these truths to our own human experiences. At the same time, this means there is no reason to expect historical accuracy or scientific details and conflicts with current science become largely unimportant. As we will see, scholars have provided good arguments that parts of Genesis were composed as symbolic stories like this.

---

11. Walton, *Lost World*, 7.

Although figurative interpretations resolve conflicts with evolutionary theory, we should proceed with caution or we may too readily dismiss real history. In just one prominent example, some may argue that Adam is a symbolic representative for the whole of mankind, but it would be a consequential error to miss the reality that such a man truly existed. Therefore, the second option is to acknowledge that early Genesis passages contain many literary techniques and figurative elements while also retaining the backdrop that they are *based on or alluding to real historical events that occurred in the past*. In other words, a story may be constructed to emphasize specific messages about a historic event without providing precise chronology or exact factual details. In contrast to pure symbolism, I will present arguments that some Genesis sections were written in this way. It is important to note that "historical" does not have to mean strictly literal in every aspect. Of course, where parts of early Genesis are based on real history we must deal with how that history harmonizes with modern scientific discoveries. The key distinctive in this approach is that it invites us to respond to the contained worldview as a real product of a historical narrative rather than merely metaphorical expressions about God and human existence.

17

Creation over Six Days

THE BIBLE PASSAGE THAT most directly provokes conflict with evolutionary theory is the creation story in Genesis 1 (Gen 1:1–2:3). This account is incredibly well-known and it is deeply imprinted into Christian culture. What can we make out in these opening verses of Scripture?

First, we observe that this is a very short and simple eyewitness account clearly devoid of subtle distinctions that might be appreciated by modern scientists. As an example, Genesis 1:11–13 testifies to God's creation of vegetation, but the writer distinguishes only (1) plants yielding seeds and (2) trees bearing fruit with seeds inside. This doesn't describe the enormous diversity of plant-like life on earth, but it undoubtedly corresponded to key plants in the minds of early Israelites concerned with food gathering and agricultural endeavors. Also, Genesis 1:14–19 notes the establishment of the sun, moon, and stars, but makes no reference to planets or other astronomical bodies that were unknown at the time.

Further, Genesis 1:20–25 states that it was by God's creative power that all animals of the earth came into existence, but the creatures are mentioned in only six broad categories: (1) great creatures of the sea, (2) other living things in water, (3) flying creatures, particularly birds, (4) livestock ("cattle"), (5) animals that move over the ground ("creeping things"), and (6) wild animals ("beasts"). The divisions are not especially valuable in the context of our modern understanding of animal types, but they must have represented a meaningful classification for the author and his audience. It is certainly understandable that the people of the day would give distinctive rank to astoundingly large sea animals, creatures with the special ability to fly, and domesticated animals upon which their lives depended.

Whatever we gather from this creation story, these verses demonstrate that the text is not intended to provide details that are comparable to modern scientific discoveries. John Calvin went so far as to acknowledge that Genesis 1:14–18 was inaccurate in implying that the moon produces its own light.[1] Given these observations, we readily understand that the author wrote within an early culture quite different from our own.

Apparent Artistry

One of the most notable aspects of Genesis 1 is its uncommon literary style that sets it apart from the straightforward narratives in most of the Old and New Testaments. In fact, this first chapter of the Bible is uniquely formatted in most translations to accommodate its distinctive character. Particularly prominent is the rhythmic pattern of verse coupling; the author expresses the same thing twice in different words. For instance, verse 1:6 reads, "And God said, 'Let there be an expanse between the waters to separate water from water." and verse 1:7 repeats, "So God made the expanse and separated the water under the expanse from the water above it." Likewise, verse 1:20 contains, "Let the water teem with living creatures . . . " while verse 1:21 responds, "So God created the great creatures of the sea and every living and moving thing with which the water teems . . . " This pattern of repetition continues through most of the chapter. The style is noteworthy because it is a literary technique more characteristic of the lyrical Psalms (poems intended to be sung) and unlike the more common historical narratives recorded in the rest of the Bible. I imagine many readers have previously recognized how distinctive Genesis 1 is in this respect.

Another apparent pattern is the use of declarative phrases inserted throughout the passage that resemble reoccurring refrains in a song or poem. These include, "And God said" (repeated ten times), "Let there be . . . " (ten times), "And it was so" (seven times), "And God saw that it was good" (seven times), as well as the recurring expression, "And there was evening, and there was morning" (six times). This crafted repetition is certainly not what we expect from an author in a straightforward telling of historical events. We may conclude that the writer is purposefully communicating through a creative, artistic style.

1. Calvin, *Commentary*, Vol 1, 1:15.

In addition, there are well-recognized, intriguing correspondences between the events described in creation days 1–3 and days 4–6 (see Table 17.1). These imply that the entire Genesis 1 account is organized into a balanced poetic framework. On day 1, God created light and distinguished "day" and "night," but it is not until day 4 that he created the sun, moon and stars to "govern the day and the night." On day 2, God "separated the water" to form the seas on the earth and the sky above, while it is later on day 5 that he created animals to inhabit the waters, the creatures of the sea and the birds of the air. Completing this pattern, it is on day 3 that dry land appeared and vegetation covered the land, while it is on day 6 that all creatures of the land, including humans, came into existence. Days 1–3 describe the creation of distinctive environments while days 4–6 describe, in perfect symmetry, the filling of those environments.

Table 17.1 — Correspondence between days 1–3 and days 4–6 in Genesis 1.

	Creation of form - environments or locations		Filling the environments or locations with inhabitants
day 1	day and night	day 4	lights to govern day and night
day 2	seas and sky	day 5	creatures of the sea, birds of the sky
day 3	dry land and vegetation	day 6	animals and humans on land

What is especially noteworthy is that the days appear to be structured as responses to the declaration of Genesis 1:2, "Now the earth was formless and empty, darkness was over the surface of the deep and the Spirit of God was hovering over the waters." Days 1–3 describe the creation of *form* from the prior formless conditions, while days 4–6 describe the creation of inhabitants to *fill* the prior emptiness or void. The overall sweep of days communicates God's creation of an ordered world. The balance between the two triads of days (1–3 and 4–6) is, again, the type of thing expected in storytelling and poetry, not in a simple detailing of the past. This suggests the inspired author arranged the creation account according to something other than a strict chronological depiction.

A straight reading of Genesis 1 also implies that the "days" of creation are not ordinary 24-hour periods. Many scholars have noted that the sun was not created until the fourth day. Since an earth day is defined by the 24-hour rotation of the earth relative to the sun, this means there would be no reference for this kind of timeframe in the first three days of the account. Likewise, the phrases, "there was evening, and there was morning" (verses 11:5, 8, and 13), that indicate the ends of these first three "day" periods, cannot refer to the normal *sun*-defined events we associate with a typical day on earth. This suggests that the creation days are figurative timeframes within the author's literary framework.

Finally, many readers have observed that, if we try to read both Genesis 1 and Genesis 2 as strictly literal accounts, there are discrepancies between the two creation stories. In particular, vegetation appears before the creation of humans in Genesis 1, while Genesis 2 indicates that God created man from the dust of the ground before any shrub or plant had appeared.[2]

Although Genesis 1 is so often understood as a literal telling of the earth's beginnings, its overall attributes indicate the author was communicating with prominent literary intentions. This is significant! Most readers approach the passage as if it was composed in the analytical, scientific style of modern Western culture. But we see that this is not the real form or content of the verses. The crucial question is to ask what this story meant to the ancient Israelites to whom it was written. This requires us to embrace a different mindset that values the richness of the story's artistic delivery.

Genesis 1 Through Ancient Eyes

As mentioned earlier, archaeological studies indicate that a key purpose of Genesis 1 was (and is) to proclaim the distinctive nature of the God of Israel in contrast to religious understandings in neighboring societies of the ancient Near East. Genesis 1 contains truth claims that serve to define a worldview by which the author and his readers approached their lives. The artistic presentation conveys these claims with a sense of grandeur and celebration. Some writers have suggested the Genesis 1 framework allowed it to be used in the context of worship where it would be read aloud, recited together or sung as a hymn. Old Testament scholar C. John Collins refers to the unique literary style as "exalted prose narrative" and he notes that this genre, " . . . points us away from ordinary narration and leads us to suppose that its proper function extends well beyond its information to the attitude it fosters."[3]

2. Kline, "It Had Not Rained."
3. Collins, *Genesis 1–4*, 44.

The profound assertions of Genesis 1 are accentuated when they are compared to surrounding belief systems.[4] Genesis 1:1 begins with the proclamation that the one sovereign God of Israel created *everything*: "In the beginning God created the heavens and earth." It is generally understood that this also means God created out of nothing ("*creatio ex nihlio*). This God alone established order out of chaos, from a formless and empty beginning on earth, and he did so to provide for humans who he made in his likeness. God's creation is inherently good, as proclaimed by God himself, and this should elicit appreciation and gladness from us. These claims contrast decisively with surrounding worldviews, such as the notions that there are many gods, that gods must wrestle with or combat material things, or that humans are either slaves to or are exposed to the whims of uncaring deities. Genesis 1 also proclaims that all physical characteristics and living creatures are God's handwork; he has authority and power over them. This stands out against religious practices that turned objects of nature into deities and included worship of the sun, moon, stars, animals or other created things. Denis Alexander provides a wonderful summary:

> This is the key text that lays down the distinctive creative character of the one true God who stands in stark contrast to the polytheistic deities so dominant in the Near East religious culture of that era. Genesis 1 is indeed foundational to the whole of the rest of Scripture, and we cannot grasp the rest of the biblical message unless we understand who it is being revealed to us in these verses.[5]

Genesis 1 and Evolution

We may now return to the question of how to effectively proclaim both the truths of Genesis 1 and the natural revelations of evolutionary biology. Many Christians, including a large number of conservative evangelicals, conclude that Genesis 1 should be understood as a metaphorical story first intended to instruct early Israelites. For example, Old Testament scholar Peter Enns states:

> The early chapters of Genesis are not a literal or scientific description of historical events but a theological statement in an

---

4. Collins, *Adam and Eve*, 137–60; Enns, *Inspiration*, 23–70; Godawa, "Biblical Creation."

5. Alexander, *Creation or Evolution*, 154.

ancient idiom, a statement about Israel's God and Israel's place in the world as God's people.[6]

In this view, there are no conflicts with evolutionary theory because Genesis 1 was never intended to convey scientific information. This approach is not allowing science to change what Scripture means; it is instead saying that the passage simply never addressed scientific matters. Importantly, reading Genesis 1 as a metaphorical story does not alter the fact that the passage communicates significant truths about God, his creation, and our relationship to him. People from thousands of years ago to today are able to understand these statements and are invited to respond. In fact, it can be argued that the value of the passage is precisely in its powerful poetic communication. Christian biologist and author Darrel Falk puts it this way:

> Indeed providing the scientific details would have obscured the real message—that of guiding people to seek God. This view also holds that the use of figurative language, with its rich symbolism and deep meaning, may have been by far the best way for God to effectively accomplish his purpose of leading his children into his Presence.[7]

On the other hand, although it is clear that Genesis 1 is not straightforward history, this does not mean the passage completely lacks implications about factual events in the past. Evangelical theologian Henri Blocher notes, "The presence of symbolic elements in the text in no way contradicts the historicity of its central core."[8] A number of scholars have argued that we should be sure to acknowledge this historical essence. At the simplest level, the statement that God created the heavens and earth is a historical claim about something that is literally true, and the succession of events in Genesis 1 implies a progression through time rather than an instantaneous creation. The key significance of these observations is that they do not stop at statements about the character of God or general truths about our lives; they acknowledge that Christians are playing a part in a true story. The story begins in the first chapter of Genesis at God's initiation, and progresses through Israel's history to the revelation of Jesus Christ, through today, and into the future.

6. Enns, *Evolution of Adam*, 56.

7. Falk, *Coming to Peace*, 32.

8. Blocher, *In the Beginning*, 155.

Creation and Time

There are different ways to see connections between Genesis 1 and literal geological and evolutionary history. It is undoubtedly wise to not push these connections too far or we risk contorting the clearly figurative account to make it say more than it does. Whatever real sequence of events might be implied in Genesis 1, most scholars would suggest the "days" of creation should be viewed as very long, overlapping periods of unspecified duration. I believe the most workable approach recognizes that God's activities are expressed through the analogy of a six-day workweek followed by a Sabbath.[9] This "literary framework" view provides a good explanation for the repeated Genesis 1 phrase that demarcates each creation day, "And there was evening, and there was morning–the nth day." That is, from the perspective of an ancient laborer, work ends in the evening and there is a time of rest through the night until the morning when the next day begins. In a analogy like this, the length of each day is not really specified and as C. John Collins explains, " . . . we can call the days and their activities broadly sequential, without requiring that every event be historically sequential; some things might be grouped by logical rather than chronological criteria."[10]

There are many observations that show we should have considerable freedom in interpreting the Genesis 1 time periods. For example, the Hebrew word *yowm* that is used for the "days" of Genesis 1 (appearing in seven different verses) is the very same word that refers to the entire creation *week* in Genesis 2:4: "This is the account of the heavens and the earth when they were created, in the *day* (*yowm*) that the Lord made the earth and heaven." We also note that God rested on the seventh day while theologians generally agree that we are still in that seventh "day." As mentioned earlier, many scholars conclude the first three days of creation cannot be typical 24-hour days because there was no sun to delineate such an earth-defined timeframe. Even further, we have Biblical reference that time is not the same to God as it is to us because of his omnipresent nature. Psalm 90:4 reads, "For a thousand years in your sight are like a day that has just gone by, or like a watch in the night." (also see 2 Pet 3:8). Old-earth creationists have been instrumental in providing good arguments like these for long, unspecified creation periods. Their contributions date back to the 1800s prior to Darwin. Although these Christians reject the full scope of evolutionary theory that I have argued for, their perspectives on long creation periods help to

9. Collins, *Genesis 1–4*, 71–83 and 122–29.
10. Ibid., 127.

solve unnecessary conflicts that some people perceive between the days of Genesis 1 and the timeframe of evolutionary history.

The Verbs for Creation

I have long been intrigued that some phrases used in Genesis 1 and 2 for the actions of creation actually imply consistency with evolution. The accounts note that the *physical features* of the universe are created from nothing with expressions like, "Then God said, 'Let there be light' and there was light" and "Let there be an expanse" (verses 1:3 and 1:6; also see Heb 11:3). However, we see something different in the creation of *living things*. For example, we read, "Let the land produce vegetation," "Let the land produce living creatures," and "The Lord God formed the man from the dust of the ground" (1:11, 1:24, and 2:7). These phrases imply that God made use of already created materials to construct plants, animals and humans. This is consistent with our modern understanding that living things are composed of the same chemical elements that are present in non-living matter, and that there is continual cycling of these elements between the two. Further, the phrasing, "let the land produce," implies that there is some mechanism or process inherent in the "land" that can bring forth plants and animals.

Particularly intriguing is the coupling of verses about living things that indicate dual involvement of natural processes and God, rather than instantaneous creation by God alone. Genesis 1:20 reads, "Let the water teem with living creatures" while the next verse speaks of the same event and says, "So God created the great creatures of the seas and every living and moving thing with which the water teems." Genesis 1:24 reads, "Let the land produce living creatures according to their kinds: livestock, creatures that move along the ground, and wild animals" while the next addresses the same events with "God made the wild animals according to their kinds, the livestock according to their kinds, and all the creatures that move along the ground according to their kinds." These coupled sentences equate the water or land bringing forth the created beings with God's creative action. Again, this implies mechanisms within the features of water and land that were part of the creative process.

I don't believe that it's difficult to see this as a representation consistent with evolutionary processes. It is especially noteworthy that the writer uses this double referencing (coupling natural processes and God's activity) for the creation of living things but not for physical things. These verses are fascinating and compelling when we look at Scripture with evolution in mind. Some authors suggest we should not interpret the text in this way because ancient readers would not have a context for the same kind of understanding. I believe this unnecessarily limits what may be contained within these God-inspired verses.

These passages illustrate how the creative and controlling hand of God can be seen in concert with natural biological processes; God created and yet the processes of evolution also brought about living things. This is a very good model for reconciling God's sovereignty with evolutionary theory. Please refer to chapter 10 for a full discussion of this important concept of dual causation.

Seeing Genesis 1 in Other Ways

There is an additional significant approach to the first chapter of Genesis compatible with the concepts of evolutionary theory. John Walton, well-respected Old Testament scholar from Wheaton College, presents an extensive analysis of Genesis 1 that he calls the "cosmic temple inauguration view.[11] The key premise is that, in the ancient world, to "create" something was not focused on the production of an object's physical properties; instead, it was about giving the object a function within a coherent ordered system. In Walton's understanding of ancient thought, Genesis 1:2 communicates that the earth was "unproductive" or "nonfunctional" (rather than "formless and empty"). The rest of Genesis 1 then describes a seven-day inauguration ceremony that is the functional beginning of the world. In days 1–3, God created by establishing functional order in the earthly spheres of *time* (light and dark, night and day), *weather* (separate waters, seas and sky) and *food* (land and vegetation). In days 4–6, God installed operating inhabitants, or "functionaries," into those spheres. Walton argues effectively that ancient peoples would readily see the seventh day as the story's climax. That is, God "rests" by taking up residence in his cosmic temple; not resting by ceasing activity, but by functioning within the ordered creation he achieved.

Walton recognizes the well-known literary correspondence between days 1–3 and 4–6, and offers good reasoning why early Israelites would have comprehended the framework as an account of functional origins, particularly functions to serve humanity. His views are well supported by his analyses of how various Hebrew terms from Genesis 1 are used elsewhere in the Old Testament. Ancient Israelites cared how the world became the meaningful place in which they lived, not how physical "stuff" came to be. In this interpretation, the days of creation are not periods of time when physical attributes of the universe were constructed (this happened before

11. Walton, *Lost World.*

the Genesis 1 account). The days are, instead, timeframes for "the inauguration of the functions and for the entrance of the presence of God . . ."[12]

A significant aspect of Walton's analysis is that it tells us that Genesis 1
does not conflict in any fashion with modern science.

> If Genesis 1 is not an account of material origins, then it offers
> no mechanism for material origins, and we may safely look to
> science to consider what it suggests for such mechanisms. We
> may find the theories proposed by scientists to be convincing
> or not, but we cannot on the basis of Genesis 1 object to any
> mechanism they offer. The theological key is that whatever sci-
> ence proposes that is deemed substantial, our response is, "Fine,
> that helps me see the handiwork of God."[13]

In other words, Walton's approach, based on biblical scholarship, allows us
to fully understand Genesis 1 while also accepting evolutionary discoveries.

A Matter of "Kinds"

I undoubtedly need to end this chapter by mentioning the repeated phrases
in Genesis 1 that indicate living things were produced or created "according to their various kinds" or "after their kind." Young-earth creationists
argue that these expressions mean biological entities remain fixed within
some type of limits and that this, in turn, precludes the possibility of large
evolutionary changes. There is really no scientific or scriptural rationale for
this position despite its common acceptance among evangelicals. In simple
observations at the time of the Genesis writer, it was clear that plants and
animals occurred in distinctive types and that they produced offspring consistent with their own type. Dogs do not have goats in their litters and fig
trees do not produce mustard seeds. The point is that God is responsible for
this order of things; he created it.

However, this has no bearing on the possibility of changes in plant and
animal form over millions of years. This concept would have been entirely
meaningless to the author of Genesis and his contemporaries. An evolutionary
view also states that organisms reproduce "after their kind." We see exactly the
same thing the biblical writer did! For what purpose, given the overall message of Genesis 1, would God be making a statement about the fixed nature of
taxonomic species, genera, or families? This is an unwarranted extrapolation
of the text. It is, at least, a very weak premise for refuting evolutionary theory.

12. Ibid., 91.
13. Ibid., 162–63.

18

Adam, Eve, and Original Sin

THE SECOND AND THIRD chapters of Genesis contain another creation account with deep significance for the Christian Church. This is the story of Adam and Eve and the Garden of Eden. There are two primary areas of contention when this passage is examined along with the conclusions of evolutionary science: First, what is the historical origin of humans on earth, including the important question, who were Adam and Eve? Second, how did humans come to possess our inherent broken or fallen condition? This condition is absolutely essential to the Christian gospel that offers the redeeming sacrifice of a crucified Savior. These are emotionally charged topics for many believers and they have become a central focus in many science-faith discussions. We would be wise to remain humble, tentative and open-minded as we work to understand what God has revealed to us.[1]

Approaching the Story Well

Genesis 2–3 is a narrative account that lacks many of the highly crafted aspects of Genesis 1 and this suggests it is more likely a historical telling. Even so, there are numerous figurative and metaphorical aspects that make a purely literal reading very implausible. Since we have learned that ancient writers often used symbolic elaboration and were not constrained by precision or chronological sequencing, we should be alert to appreciate creative messages.

---

1. See van den Toren, "Not All Doctrines."

For example, in Genesis 2:19–20 God brings "all the wild animals" and "all the birds of the sky" to Adam so that he can provide each a name. Of course, it stretches credibility that God paraded before Adam every animal type from around the entire earth. On the other hand, this depiction communicates an orderly creation orchestrated by God in which animals exist in distinct recognizable types. Also, the act of "naming" in the ancient world represented control or dominion over the named object. In other words, ancient readers would understand that the world was created for their benefit and was theirs to rule.

A number of interpreters have noted the Garden of Eden itself is a symbolic representation for a sanctuary or temple.[2] For example, Israel's temple was where God's unique presence rested in much the same way as Genesis describes God occupying the Garden. Adam was also brought to the Garden "to work and take care of it" in a fashion parallel to priests caring for the temple. Significantly, the two Hebrew words translated from Genesis 2:15 as "work" (abad) and "care" (samar) are most often used together elsewhere in the Old Testament for the role of priests who "serve" and "guard" the temple.

Ancient Israelites would also understand that God's earthly blessings and provisions emanate from his temple. Genesis 2:10–14 describes a river flowing out of Eden that separates into the *headwaters* of four major rivers. In normal geography, a single river flows downhill and merges with other tributaries. In notable contrast, a real river does not diverge to form the beginnings of four different primary watercourses! A symbolic reading is particularly meaningful when we note that two of the four mentioned rivers, the Tigris and Euphrates, are the main, life-giving, waterways of the Mesopotamian region where Genesis was written (the other two rivers are unknown).

Some scholars have pointed to the formation of Eve from Adam's side to be his helpmate (Gen 2:21–22) as a rich metaphor for the equality of women and God's relational plan for humans in marriage.[3] Whether or not there was an actual physical operation on Adam's body may be of minor significance compared to these messages. Another well-recognized symbolism is the use of the Hebrew word "*adam*" for the first man of Genesis 2. This is a term that means all humans collectively, as well as being an individual's name. What better way to tell the story of human beginnings than to give the star of the tale a name with this perfect double meaning?

Finally, the story of "The Fall" in Genesis 3 appears particularly full of imagery. Examples include seemingly figurative references to the "tree of

---

2. Alexander, *Creation or Evolution*, 257–58; Beale, "Eden, The Temple"; Beale, *Temple*; Lioy, *Axis of Glory*, 5–16.

3. Alexander, *Creation or Evolution*, 196–98.

life" and the "tree of the knowledge of good and evil," as well as to a "crafty" talking serpent. Once again, these features do not mean the account lacks a historical basis, but we should clearly be open to literary meanings intended by the inspired author. With this framework, we may proceed deeper into the Adam and Eve stories.

The First Humans

One of the most troubling aspects of evolutionary thinking for many Christians and non-Christians alike is its implications about the origins of humans and our comparative position among all living things. Our hearts and minds tell us that surely we have a unique place in this world. The Bible gives an emphatic affirmation of this sense. Scripture teaches that humans, both male and female, were created in the image of God and that we were always meant to have a distinctive existence dependent on a relationship with him (Deut 8:3). God cares for us above all other aspects of his creation (e.g., Matt 10:29–31). God's ultimate love for us and offer for intimacy and fulfillment are extended through Jesus (e.g., 1 John 4:9, Titus 3:4–7, and Rom 5:8).

"This is all good," you might say, "but what about the notion that humans evolved from apes? I simply can't buy that!" I have heard this expression or something similar an uncountable number of times. The message is pointedly that no matter what reasoning I might present this is a non-negotiable issue. I understand the disturbing nature of the discussion. For many people, the idea that we share an ancestry with apes is insulting or disheartening, and it reduces humans to nothing more than animals without uniqueness.

However, we are faced with a number of profound observations. Humans and great apes (chimpanzees and gorillas) are actually very similar in physical form and internal physiology and possess DNA and other molecules that are more similar to one another than to any other creatures. To an unbiased eye, the comparison of humans to apes is not unlike matching dogs with foxes or deer with elk. Nobody would deny the apparent connection between the latter pairings, but we find it difficult to see the former grouping in the same way. But, in fact, humans and great apes are *more* similar to one another in their DNA than the species in the other pairings. Further, the discovery of a variety of fossil hominid species (e.g., *Australopithecus*, *Homo* sp.), which appear to fall between great apes and humans, brings the similarity of humans to other species even closer.

Humans also share uncountable chemical and biological features that are universal to all living organisms. These include the use of the same classes of organic molecules, genetic instructions encoded in the same way on DNA molecules, the same protein synthesis mechanisms, dependence on the same ATP energy-carrying molecules, and much more. Humans are unremarkably mammals by all definitions; including mammalian hair, teeth, heart, blood vessels, digestive tract, reproductive systems, and more, as well as milk-producing mammary glands in nursing mothers. From a biological view of comparative anatomy and physiology, there is really nothing conspicuously distinctive about humans compared to other organisms. The most striking observation is not at all our physical uniqueness; rather, it is the physical unity of humans with all living things! We insist that we are unique but our biology provides no evidence for such a claim.

I suggest that our rejection of human relatedness to other species is more an emotional response than a reasoned one. Something tells us that because of our special spiritual nature we must also be singularly different and unique in *physical form*. But Scripture does not specifically say this. Instead, Genesis states that both humans and animals were formed "out of the ground" (Gen 2:7, 2:9) with no indication that humans and other organisms were created by different methods. If God created humans without evolutionary steps, why are we so fully similar to other living forms, why do we share a particular relatedness to mammals, and why do we bear such a remarkable likeness to great apes? Is it not possible that our insistence of physical uniqueness comes from an inflated view of ourselves (certainly a common biblical theme, e.g., Rom 12:3)? The most consistent interpretation is that the human form prior to Adam evolved through evolutionary ancestral connections. If we proceed with this conclusion, how can we read the biblical story to affirm the profound uniqueness of humans and simultaneously acknowledge these scientific implications?

I believe the answer revolves around what it means that God created humans "in His own image." Few people would argue that the God of the universe has human arms, legs, and facial features. Our human distinction is not in physical form but in other qualities that God himself possesses. We are made like God in our ability to reason and understand and form a spiritual relationship with our Creator. We alone among living things are able to hear and respond to the word of God. We are created to have fellowship with him (1 John 1:3), to love him with all our heart, soul, strength, and mind (Deut 6:5, Luke 10:27), and to worship him (Deut 6:13; Ps 34:1, Ps 95, Ps 103, and many other psalms). We also have unique potential for knowledge (Col 3:10) and spiritual attributes such as righteousness and holiness (Eph 4:14). With these characteristics, humans are distinctly created to be God's representatives on earth, as stewards over the rest of creation. This unique makeup is a special "image" character that God gave to humans that is absent from other creatures. As "image-bearers," we are utterly unique and significant to God.

We have good reason to conclude that the human physical form is connected to all other living things through evolutionary relationships, but also that we bear a distinctive "image' character. How, then, may we approach the Genesis 2–3 account?

Who Were Adam and Eve?

For some Christians, Adam and Eve are powerful symbols for the beginning of humanity and the Genesis chapters use this first human couple as representatives in a meaningful metaphorical account.[4] Dan Harlow, biblical researcher and professor at Calvin College, expresses the view succinctly:

> Adam and Eve are strictly literary figures—characters in a divinely inspired story about the imagined past that intends to teach primarily theological, not historical, truths about God, creation, and humanity.[5]

From this perspective, Genesis provides a description of who humans *are*, and what they have always been. The passage proclaims and celebrates that humans were created as a focus of God's activities, that they were intended for a unique role in relationship with him and to be his representatives on earth. A metaphorical understanding suggests there are no historical descriptions in the Adam and Eve story and this means there is no conflict with the evolutionary development of humans.

Within this thinking the origin of special "image" characteristics in humans can be approached in different ways. Some authors suggest that God directed human evolution toward unique mental capacities for self-awareness, the ability to distinguish moral differences, to understand the possibility of a living God (or gods), and to choose whether to follow Him.[6] Others visualize that, even if Adam and Eve did not exist, God did something extraordinary apart from evolution to affect the earliest humans, to impart a new kind of spiritual perception.[7] One thing to keep in mind is that details about the entrance of God's image into humanity will never be revealed in fossil evidence or in the structure of DNA.

For conservative Christians who are inclined to object, you should know there is considerable scholarship to support this position. Although

---

4. Enns, *Evolution of Adam*; Harlow, "After Adam"; Lamoureux, "Was Adam."

5. Harlow, "After Adam," 181.

6. Collins, "Original Sin"; Day, "Adam."

7. Collins, *Language of God*, 206–10.

this approach may challenge previously held views of Genesis 2, we should be ready to abandon old thinking if it contains incorrectly attached history or science that was never intended in the original writing. One thing we should openly accept is that a metaphorical reading conveys rich and compelling truths.

On the other hand, many Christians are unsatisfied with the proposition that Adam and Eve are not real individuals in history and many would insist that at least some aspects of the Genesis account are real events. I am among them. If we hold to the historicity of Adam, we are faced with explaining how Adam can be understood as a wholly unique human figure that also evolved physically from hominid ancestry.

One option, which I embrace, is that Adam was singly taken aside by God from physically evolved humans and the image of God was divinely imparted to him.[8] Adam immediately entered into a relationship with God. In other words, Christian believers may accept that the origin of the first image-bearing human was an extraordinary event in which God chose the evolved physical form of Adam and breathed a unique spirit life into him. A key distinction in this view is that God specially bestowed this image onto Adam and it has been passed on to offspring; it was not something that simply evolved along with human physical features. This approach harmonizes the biblical account of Adam and evidence for the evolutionary development of humans.

A key challenge to this view is that detailed DNA studies do not indicate that modern humans derived from a single originating pair.[9] Several lines of genetic evidence suggest that we are all related to a group of humans of at least several thousand individuals (a shared ancestry that is significant in itself) but that we did not all come from two unique parents. One well-discussed suggestion is that God's image was first bestowed to Adam and then was subsequently bestowed to Adam's related contemporaries.[10] This conceptualization of Adam as the "federal" head of humanity is consistent with common biblical themes in which the characteristics and actions of a singular individual (generally, a king) are imparted to and affect those he represents. A population of humans that were joined into the uniqueness of Adam lines up with the scientific data. For those who might think that this is an invented suggestion in response to current advances in genetics, it was first proposed in 1967 before modern studies of the human genome.[11]

8. Kidner, *Genesis*, 28; McIntyre, "Historical Adam"; Schroeder, *Science of God*, 131–52.

9. Falk, "Does Genetics"; Venema, "Genesis"; Venema, "Understanding Evolution."

10. Collins, "Adam and Eve as Historical People," 160–61; Keller, "Christian Laypeople," 10–12.

11. Kidner, *Genesis*, 26–31.

A parallel but somewhat different proposal suggests that God revealed himself in a special way to two individuals or a group of humans and this knowledge of God spread outward to other people who would hear.[12] The idea is that the original couple or community was chosen from among other existing humans to know God personally and represent him as stewards on earth. In this approach, Adam and Eve are real individuals and the federal heads of God's image. What distinguishes this model from the one above is that God did not specially impart new characteristics (an image) to humans; rather, God divinely revealed himself and offered relationship to people who already had the capabilities to respond. Other humans joined by choosing to participate, not through genetic descent from the first image-bearing humans.

For those who insist on the historical reality of Adam and Eve, we must accept that there will always be mysteries within the Genesis narratives concerning the first "image-bearers." For example, after Adam and Eve were banished from the Garden of Eden, their sons, Cain and Seth, had wives with which they had many offspring. Where did these wives come from? The traditional argument is that these were sisters born from Adam and Eve,[13] but this is not obvious or mentioned directly in the biblical text. Others suggest that it is unlikely the wives were sisters given the later strong Biblical prohibitions against such unions.[14] Also, who were the people Cain was afraid would kill him when he was driven away to wander the land (Gen 4:17)? The story line strongly implies there were a fair number of other humans around with which Cain intermingled after he traveled far from his parents. The point is that it is unwarranted for anyone to claim with certainty that he or she knows historical answers to such questions.

Where Did Human Sinfulness Come From?

The most critical challenge from evolutionary biology for many Christians, especially evangelicals, is its conflict with a traditional view of the fall of humanity through the actions of Adam and Eve. The issue is not simply about the validity of an ancient story. Instead, it is about the core soundness of the New Testament and the authenticity of Jesus as the incarnate Savior. Without the foundational notion of the sinfulness of humans, there is no corresponding need for salvation and restoration later provided by God through Jesus' death and resurrection. This is at the very heart of Christian faith.

For some believers, the story of "The Fall" in Genesis 3 is best understood as a comprehensive metaphor for our inherent human condition and

12. Alexander, *Creation or Evolution*, 214–42; Hurd, "Hominids," 222–28.

13. Batten, *Creation Answers*, 127–39; Ross, *Genesis Question*, 101–06.

14. Collins, *Language of God*, 207; see Lev 18:9 and 20:17, Deut 27:22.

real tendencies in relation to God.[15] In the fall of Adam and Eve, we see a clear expression of the human desire for knowledge, power, and control. We see the symbolism that this nature leads to separation from God and both spiritual and physical harm. This state is contrasted with a place of peace and harmony where we are in a dependent and submissive relationship with the One who formed us. In this view, the point of the account is not the historic origin of our human condition, but to provide a vivid word picture of its character.

Christians acknowledge that we repeat the story of Adam and Eve in our own lives on a regular basis. We are tempted and doubt that God's path for us is best or we acknowledge the good we want to do or ought to do, but we do not always follow through (see Rom 7:14–25). Our tendency is to rebel against what we know is good and this leads to separation from a right relationship with our Creator. To Christian believers, it is through faith in and dependence on Jesus that this relationship is restored. This is sound theology and, for many individuals, a figurative reading is a fully satisfying rendering of the Genesis passage. As Christian scholar Denis Lamoureux concludes, "Adam never existed, and this fact has no impact whatsoever on the foundational beliefs of Christianity."[16]

The many figurative aspects of the Garden of Eden story (e.g., trees of life and knowledge of good and evil, talking serpent) provide support for this metaphorical approach. Authors have also pointed out that Adam is not directly referred to as the cause of sin and death in the Old Testament after Genesis 3 and they suggest the real emphasis is the universal reality of human sinfulness, not the origin of it.[17]

However, there are a number of challenges in reading the fall of Adam and Eve entirely as metaphor and many commentators (and probably most Christians) accept that some aspects of the account are real events in human history.[18] A primary issue is that Adam is described in the New Testament as a real person who specifically brought spiritual and physical death to humans. The apostle Paul develops key foundations of Christian faith by directly pairing the fall of humanity through Adam with the salvation provided through Jesus. Especially significant passages are Romans 5:12–21 and 1 Corinthians 15:12–22. Consider the following verses:

15. Collins, "Original Sin"; Enns, *Evolution of Adam*; Lamoureux, *Evolutionary Creation*; Murphy, "Roads to Paradise."

16. Lamoureux, *Evolutionary Creation*, 367.

17. Enns, *Evolution of Adam*, 82–84.

18. Alexander, *Creation or Evolution*, 214–42; Collins, *Adam and Eve*; Keller, "Christian Laypeople," 7–12; Walton, *Lost World*, 68–70.

> For if, by the trespass of the one man, death reigned through that one man, how much more will those who receive God's abundant provision of grace and of the gift of righteousness reign in life through the one man, Jesus Christ. Consequently, just as the result of one trespass was condemnation for all men, so also the result of one act of righteousness was justification that brings life for all men. For just as through the disobedience of the one man the many were made sinners, so also through the obedience of the one man the many will be made righteous. (Rom 5:17–19)

> For since death came through a man, the resurrection of the dead comes through a man. For as in Adam all die, so in Christ all will be made alive. (1 Cor 15:21–22)

Paul is unquestionably referring to Adam and his actions as if they are a historical reality on a par with the reality of Jesus.

Those who interpret the Adam and Eve accounts as metaphor respond that later commentators have misunderstood Paul's use of Genesis 2–3.[19] They avidly agree that what Paul says about Adam is a crucial aspect of Christian theology, but also contend that a figurative Adam does not harm the core truths of Paul's communication. Some authors suggest that Paul and his Jewish contemporaries mistakenly believed in a real Adam (just as they had inaccurate scientific views of the world) and that God worked with these common perceptions when he inspired Paul. Others suggest that Paul knew Adam was not the cause of "original sin" but crafted a history-like analogy to express his key point of salvation through Jesus.

Both approaches to a symbolic Adam carry problematic implications. First, they certainly weaken the obviously powerful connection Paul is making between a singular *real* person in Adam and the singular *real* entity of Jesus. Second, they make us wonder about the inspired authorship of Paul, the primary writer of the New Testament letters. It is one thing for Paul to have inaccurate perceptions about natural science that are not corrected in Scripture, but it is something else for him to be wrong or figurative about real world descriptions that are so central to his theology. If Paul is wrong or loose with the historic reality of Adam, is he credible in his interpretation of events in his own time or how he saw Jesus in other passages of the Old Testament?

A common biblical expectation in evangelical churches is that the entirety of the Bible points to Christ. We see prophetic references to Jesus and assume the inspired authors did not fully understand what they were writing because the text is ultimately the Word of God. Paul models this

19. Barr, *Garden of Eden*; Enns, *Evolution of Adam*; Harlow, "After Adam."

approach when he refers back to the Old Testament to see and proclaim Jesus as the Messiah. But if Paul is wrong or purposely artistic about Adam, isn't he just as likely to be inaccurate or unclear in his other insights? The profound implication is that this challenges the inspired nature of Scripture and its authority. It tends to reduce the Bible to a human-constructed book containing inaccurate connections and/or poetic manipulations that are difficult to decipher.

Another difficulty with reading The Fall as metaphor is it is inconsistent with the core biblical themes of redemption and restoration. These concepts are expressed repetitively in the Old Testament with regard to the nation of Israel, and are later applied to the saving work of Jesus in the New Testament.[20] For example, we read in Romans 3:22–24: "There is no difference between Jew and Gentile, for all have sinned and fall short of the glory of God, and all are justified freely by his grace through the redemption that came by Christ Jesus." Redemption and restoration are terms that primarily mean a return to a state that once existed but was lost. Importantly, they do not typically mean improvement or correction from an inherent negative condition that was always present. Paul's message (and the message throughout Scripture) makes most sense if Jesus redeems us to a condition once experienced in Adam. This fundamental Christian proclamation is muddled if human sinfulness has always been an aspect of human nature.[21]

Christian believers readily acknowledge that the account of Adam and Eve is a most unusual situation. However, for those who have faith in a sovereign Creator-God and accept other divine miracles expressed in the Bible, it is reasonable that God would establish an entirely unique setting for the origin of the first humans that he created as image-bearers. If this is so, how do historical aspects of the story fit with evolutionary biology? We may actually discern many things from the biblical verses themselves.

Pain and Death in God's Creation

One unworkable interpretation of Genesis 2–3 is to insist that all suffering and death entered the world through the actions of Adam and Eve, and that such things were entirely absent from the world before their disobedience. This reading would preclude a long evolutionary history of organisms prior to this first human couple. I provided a number of biological and biblical reasons

---

20. e.g., Ps 111:9, Ps 130:7–8, Isa 1:26, Acts 3:21, Eph 1:7, Col 1:13–14.

21. Denis Alexander concludes, "The heart of the biblical doctrine of the Fall is about relationship with God that was broken . . . A relationship cannot be broken by sin unless the relationship existed in the first place." *Creation or Evolution*, 274.

in Chapter 5 as to why this assumption cannot be accurate. For example, the Genesis passage does not describe a Garden that was a perfect paradise and instead indicates a pleasing setting that Adam had "to work" and "take care of." Also, a carnivorous animal like a lion, with its form for capturing live prey and its teeth and digestive tract adapted for meat, is a meaningless construct without the suffering and death of other animals as its food. There is simply no way for a lion to survive eating plant material (which requires plant death, in any case). It was further noted that Eve's pain in childbirth was *increased* after banishment from the Garden (Gen 3:16) rather than it beginning at that point. What does the story of The Fall really tell us?

Especially important information comes from Paul's very words in Romans 5. A full reading of this chapter reveals that its primary message is that humanity has lived in a state of *spiritual death* since the time of Adam. This spiritual death is our separation from God. Just as Adam was cast out of the Garden from a place of harmony and relationship with God, our own human nature separates us from a right communion with him. Adam did not physically die from his act of disobedience; rather he experienced the condemnation of being banished from God's presence. Also, the point in Romans 5 is that Jesus' act of righteousness "brings life" to those of us who are already biologically living. Through Jesus, humans are offered grace and a new opportunity to be restored into a personal relationship. Paul refers to death and life in the same way in Romans 8:6: "The mind of the sinful man is death, but the mind controlled by the Spirit is life and peace." If the point of The Fall is about human spiritual death or separation from God, then use of Genesis 3 to object to evolutionary processes is unfounded.

However, 1 Corinthians 15 seems more likely to be connecting the fall of Adam to *physical death*. Paul's main discussion point in this chapter is the reality of Jesus' resurrection from a state of physical death. In fact, Paul states, "And if Christ has not been raised, our preaching is useless and so is your faith" (1 Cor 15:14). It is in this context of resurrection that he writes "death came through a man" and "in Adam all die." Is it possible that all physical death started with the fall of Adam after all?

Many commentators have noted that the answer lies within the story of Adam and Eve itself. Intriguingly, nowhere does the Bible actually imply that Adam was ever free from a destiny of physical death or that he possessed eternal life. Read Genesis 2–3 carefully and one discovers that the real communication is that the *possibility* of eternal life was present in the Garden in the tree of life, but Adam and Eve never obtained it. In fact, after Adam and Eve ate from the tree of the knowledge of good and evil, God said, "The man has now become like one of us, knowing good and evil. He must not be allowed to reach out his hand and take also from the tree of life and eat, and live forever" (Gen 3:22). This shows that the possibility for eternal life was a special circumstance in the tree of life that was never realized. Access back to the "tree of life" through Christ is beautifully expressed in Revelations 22:14: "Blessed are those who wash their robe, that they may have the right to the tree of life and go through the gates into the city."

The logical corollary is that Adam's inherent condition was he would die a physical death. Again, there is no indication in Genesis 3 or in any New Testament passage that this inherent physical nature was altered as a result of The Fall. This interpretation is not new and, in fact, it dates back at least to Augustine. The imagery of an immortal Adam in a perfect Garden is not scripturally sound and it has been unnecessarily propagated. We can conclude that the physical death experienced by all humans follows from, or came through, Adam's actions. But the event was a *lost potential for eternal life later restored to believers through Jesus*. It was not an eternal, perfect Adam being plunged into an new state of physical decay and ultimate death. These are very important and different things.

There is also no reason to expect that the unique tree of life in the Garden, which God placed there in the context of the first image-bearing humans, was eaten by any other creatures or obtained by plants (whatever that might mean). In other words, we can assume that normal biological processes, such as photosynthesis by plants and digestion by animals, as well as growth, reproduction, decay, and death, occurred continuously from long before Adam and Eve to the present day. Also, since Adam was not inherently immortal, there is no reason to expect that other living things were. This means that the account of The Fall in Genesis does not conflict with a long evolutionary history of living and dying organisms prior to the Garden of Eden.

One particular passage in Genesis has been especially implicated as a problem for an evolutionary view. Genesis 3:17–18 records God's pronouncement as Adam is cast out of the Garden:

> Cursed is the ground because of you; through painful toil you will eat of it all the days of your life. It will produce thorns and thistles for you, and you will eat the plants of the field. By the sweat of your brow you will eat your food until you return to the ground, since from it you were taken; for dust you are and to dust you will return.

Many Christians insist these verses mean God issued a curse that enacted a fundamental change in the natural processes of his created world. But it is unlikely that this is what the passage is communicating. We have learned it is a mistake to look for descriptions of natural mechanics in ancient theological literature. Surely, the primary message is that Adam (and other humans after him) was cursed to live a life with increased toil and pain compared to that which he would have experienced under direct care and communion with God in the Garden. It is significant that the curse refers only to the *ground*, not the entirety of creation, and stresses the challenging agricultural work Adam and his descendants would have to face to sustain themselves. The announcement that Adam would "return to the ground" is in perfect accord with the discussion above; Adam lost the opportunity for eternal life and would die a normal physical death. What is most important, these verses do not state that all suffering and death entered the world at this time. This means they are not a strong source for negating evolutionary theory.

Finally, there is a passage in Romans 8:18–25 that has challenged theologians for centuries. In particular, Romans 8:19–21 reads:

> For the creation waits in eager expectation for the children of God to be revealed. For the creation was subjected to frustration, not by its own choice, but by the will of the one who subjected it, in hope that the creation itself will be liberated from its bondage to decay and brought into the freedom and glory of the children of God.

Whatever Paul is referring to, his focus is on the future glory that will be revealed when the children of God are fully redeemed. The verses state that creation experiences negative consequences because of "the will of the one who subjected it," the latter presumably being Adam. The word translated as "decay" implies that physical degradation came through the actions of Adam. But this is undoubtedly based on a presupposed (and unnecessary) interpretation of Genesis 3. The same word may be translated "corruption" (as it is in the ESV Bible), which emphasizes dishonest and depraved human behavior (see Gen 6:11–12). This means the natural world is currently subject to the sinful, corrupt actions of fallen humans, not some dysfunction in the normal mechanisms of nature.[22] Humans, who are out of communion with God, have certainly enacted disastrous stewardship over the natural creation; they interfere with the beauty and glory of God that nature proclaims and they dishonor the covenant that God himself made with the natural world (see Gen 9:12–13). In other words, Paul is correct that the creation suffers from corruption and it will be redeemed along with the children of God!

Some Christians undoubtedly will be uncomfortable with the previous paragraphs because of already established beliefs. This brings us back to the

---

22. Alexander, *Creation or Evolution*, 268–69; Collins, *Genesis 1–4*, 182–84.

overarching purpose of this chapter. The natural world presents abundant evidence for a long and complex evolutionary history. I've urged that this information must be allowed to inform us how to correctly interpret biblical passages, and this is what I have done. It is up to Christians to discover meanings in these verses that celebrate and integrate the totality of God's revelations to us as we seek to better know the Creator-God whom we worship.

A Summary

The Christian church has been challenged throughout its history by how to properly interpret and apply the Genesis account of Adam and Eve. I certainly have not addressed all questions that may be raised about its meanings. Instead, I have offered a sampling of alternative ways to think about its significant messages in light of modern biology.

In one approach, the story of Adam and Eve in the Garden of Eden can be interpreted as a rich metaphorical representation for humanity's unique status in relation to God, our inherent human tendencies for control and self-enhancement, and how our weakness in doing what is good damages a right communion with him. This understanding is easily compatible with an evolutionary view of the origin of humans.

Still there are important reasons to expect that the Genesis passage of Adam and Eve communicates some real aspects of history in which God interacted with the first humans in an extraordinary fashion. Evolutionary biology is in harmony with this view if we see that the key attribute of Adam and Eve is their spirit nature, the "image of God." At the same time, the physical form of Adam, and all humans, is remarkably united with other living things, indicating in every way that it is the product of an evolutionary past.

The wholly unique events comprising "The Fall" of Adam and Eve are also compatible with evolution when we appropriately recognize what they imply about spiritual and physical suffering and death. While Adam's fall established spiritual separation from God and, in turn, assured physical death, the Genesis passage does not indicate that suffering and death for plants, animals, or Adam and Eve were absent prior to the event.

Most important, the discussion in this chapter demonstrates that individuals with a vital and engaged Christian faith may openly dialogue about the intersection of the Bible's meanings and the evolutionary message in God's natural creation. The point for all Christians is not at all to water down or compromise our beliefs, but to grow a deeper, excited and informed relationship with the risen Lord in the midst of a scientific age.

19

The Biblical Flood

ONE OTHER SECTION OF Genesis has had special significance in the history of creation-evolution debates. This is the story of Noah's ark and the great flood in Genesis 6–8. A well-known depiction is a worldwide inundation with waters covering the earth to a depth of 20 feet above the highest mountain peaks, and Noah's ark filled with a pair of every land-dwelling animal species. This account is popularly taught to Christian adults and children and it is embraced as a real historical occurrence within many evangelical churches. Does this storyline mesh with God's natural revelation?

There are numerous reasons to doubt that an event of such enormous proportions ever took place. First, as with earlier Genesis chapters, literary attributes of the text suggest that it is not a straightforward telling of history. For example, the flood narrative is organized into a "chiastic" structure where elements in the first and second halves of the story are mirror images of each other.[1] One example is that the waters increase for forty days to cover the land during the first part of the story, matched by forty days for the water to recede in the second half. Second, as Western cultures accumulated full knowledge of the massive size of our planet, and the vast array of animal types on earth, many Christians correspondingly recognized that the Genesis flood probably did not communicate a global event. Later geological and evolutionary discoveries have further indicated that a worldwide inundation is highly unlikely.

---

1. Lamoureux, *Evolutionary Creation*, 216–20.

Most Christians scholars conclude that the narrative of Noah's flood is either an entirely figurative story with important symbolism or a creative account about a significant event in a localized geographical area. In support of the latter view, there are written narratives of a terrible flood from other ancient Near East cultures surrounding the early Israelites.[2] Many old-earth creationists and evolutionary creationists comfortably accept that Genesis 6–8 is based on a historic flood of local extent. Importantly, a good number of very conservative evangelicals acknowledge the merits of this interpretation.[3] Even so, unwavering young-earth creationists continue to defend a worldwide inundation.

A simple Google search reveals an extensive array of writings that attempt to explain the worldwide flood position, as well as a comparable assortment of articles and books that refute it. I do not wish to offend fellow brothers and sisters in Christ, but I believe this should be a nonnegotiable topic among Christians who seriously accept what is revealed in the natural world. A global flood viewpoint is a clear example of developing natural science by reading the Bible alone and then searching the earth for evidence to support what has already been decided. I will cover only a sampling of topics in this chapter to illustrate some central issues.

All believers may keep in mind that the primary theological messages of Genesis 6–8 remain the same whether the flood event is metaphorical, localized or global. In all interpretations, significant lessons are communicated about human tendencies toward corruption verses faithful perseverance, and God's corresponding actions of judgment and grace.

Every Living Thing

A particularly confounding issue has been how readers translate seemingly all-encompassing phrases in the flood account. These include "every living thing on the face of the earth" (7:23) and "under the entire heavens" (7:19). These sweeping descriptions understandably cause modern readers to visualize a global event. Once again, the cultural context of the written words is important. The author's use of these phrases can be understood in two ways. First, they may be exaggerated expressions intended to emphasize the powerful message of a metaphorical account. Alternatively, if the flood is a real event, the phrases are best understood to mean the entire *known world*

---

2. Ibid., 220–23; Collins, *Adam and Eve*, 143–46; Enns, *Evolution of Adam*, 46–50.

3. A localized flood interpretation is offered as an option in the *NIV Study Bible*, 15. This is a well-accepted and utilized Bible translation in conservative evangelical churches.

of the author and his contemporaries. The known world is always the frame of reference in the Old and New Testaments for universal terms like this, not the entire earthly sphere of which people had no knowledge.

Various observations argue for a limited scope. The term translated as "earth" in the flood narrative (*erets*) also means simply land, ground, or a specific geographic region or country. It is translated like this in the rest of Genesis. In the same way, the term for "heaven" (*shamayim*) translates accurately as "sky." This means that presumed global references literally mean "every living thing on the face of the land" and "under the sky." Later English Bible translators (particularly those for the King James Bible) thought that the events of Noah and his ark should be imparted as a worldwide story, so they chose English phrases to communicate this global viewpoint. But this is not clearly implied in the original Hebrew.

There are many Bible passages that use all-encompassing descriptions when the real meaning was a localized area. Genesis 41 describes a famine in the lands around Egypt while Joseph was in the service of Pharaoh. The text reads, "And *all* the countries came to Egypt to buy grain from Joseph, because the famine was severe in *all the world*" (41:57) [my emphasis]. No one understands this to mean that the famine extended to North America, South Africa, and the Far East or that people traveled from these distant locations to buy grain from Joseph. Instead, we recognize that "all the world" meant to the author only the far-ranging lands that were relevant within Egypt's influence.

Similarly, 1 Kings 10:24 proclaims, "The *whole world* sought audience with Solomon to hear the wisdom God put in his heart" [my emphasis]. No one would argue that people came to see Solomon from distant continents that were beyond the knowledge of the Israelites. Even in the New Testament, Paul claims that the faith of believers in Rome was being reported "all over the world" (Rom 1:8) and that the gospel was bearing fruit "all over the world" (Col 1:6). The universal world to Paul at the time was primarily the lands of the Roman Empire. We must allow biblical authors to express themselves with the perceptions and language of their time without forcing an absolute literalism based on our modern-day understandings.

If the flood is understood as a real event in a specific geographic region, it can be placed in the area around the Euphrates and Tigris rivers in what is now Iraq. This is a floodplain surrounded largely by low mountain ranges, including the extensive mountains of Ararat among which the Bible indicates the ark landed. Identifiable named places in Genesis through the time of the flood are in this geographical area and this may well have represented the known or relevant world for Noah at the time.

In a local flood interpretation, many presumed global references in Genesis 6–8 are easily understood to signify significant but smaller-scale indications. For example, verses that state that the mountains were covered in water mean the relatively low, visible mountains around the Euphrates/Tigris floodplain rather than all the incredibly high peaks scattered around the globe. Also, although tradition has it that Noah's ark came to rest on Mt. Ararat (which reaches to an almost 17,000-foot elevation), the Genesis text has *never* indicated this. Instead, the ark came to rest "on the mountains of Ararat," which simply means somewhere within the primarily low Ararat mountain ranges (Gen 8:4).

A regional flood translation eliminates enormous difficulties inherent in a worldwide scenario. In particular, how could Noah load animals from around the entire earth sphere onto his ark? The local flood view suggests that only significant land animals from a limited region of the Near East were ark passengers. It is noteworthy that the animals God instructs Noah to carry (Gen 7:14) are only four of the six broad categories of created animals described in Genesis 1. The excluded groups are the water-dwelling creatures that presumably did not need to be brought onto the vessel. As in Genesis 1, the emphasis is on animals significant to the people of the day and the verses can be interpreted as overlooking many animals, especially lower forms such as snails, worms, spiders, and insects. All of this suggests that the listing of animals on Noah's ark should be understood as figurative expressions for important creatures, rather than an inventory with scientific significance.

To be sure, whatever aspects of Noah's ark are real rather than metaphorical, they imply God's intervention for the loading and survival of animals, including the gathering of creatures to enter the boat (see Gen 7:8–9). However, a regional flood avoids unfathomable phenomena that are not required by the biblical text. One can always insist that anything is possible with God, but the notion that God somehow caused animals to migrate from every part of the earth, by unknown methods, makes the global flood story so miraculous that the focal work of Noah and the existence of his ark in the Near East seem almost inconsequential or irrelevant by comparison. There is no essential theological rationale for why Christians must accept an extremely unlikely global event.

Modern Arguments for a Global Flood

Despite good reasons to doubt a worldwide immersion, many evangelicals still hold to a global deluge. Flood conceptualizations became formalized in 1923 with George McCready Price's *The New Geology* and were more fully detailed in 1961 by John Whitcomb and Henry Morris in *The Genesis Flood*. These books attempt to explain how a worldwide flood only several thousand years ago can account for almost all geological features observable on the planet today. This includes monumental features such as sedimentary deposits that are many miles thick, a great diversity of plant and animal fossils, huge fossil-fuel reservoirs, massive seafloor spreading, uplifted mountains, volcanism, mountain erosion, and the drifting of continents. These books also offer explanations for the specific animals Noah took on the ark. The contained arguments have been maintained and extended by current young-earth organizations.

The tenets of "flood geology" are not always outwardly included in presentations of young-earth creation. However, they are an absolutely essential element of *any* young-earth position. Without a global flood in the recent past it is impossible to explain how a very young planet displays such enormous depths of sedimentary deposits and the fossilized remains they contain. All other explanations imply a very long and complex earth history. The fossil record distributed through the sediments is also very strong evidence for evolutionary theory. Young-earth creationist leaders have understood these constraints and this explains why flood geology is so crucial to their arguments.

However, to put it frankly, incredible calisthenics are required to hold together a worldwide flood construct and there is virtually no scientific support for it. It may seem sensible at first that a global submersion could lay down all the sedimentary rock around the earth, but there is no reasonable way that this type of flood can account for the *sequencing* of animal and plant fossils through adjacent layers. For example, we know that mammals and bird fossils occur only in relatively recent sediments and fish alone among vertebrates occupy older, deeper layers.

Flood geologists acknowledge the notable sequencing in fossil strata but they argue that it's due to a variety of factors related to a worldwide flood.[4] One suggestion is that different locomotory abilities of animal types determined where each animal form was laid down. However, it is difficult to understand how this can explain small mammals, such as rodents, shrews, and bats, *always* occurring in more recent sediments compared to extinct flying reptiles (pterosaurs), large amphibians that could swim (labyrinthodonts), and extinct dolphin-like reptiles of the dinosaur era (ichthyosaurs). And why didn't a single dinosaur survive long enough to be deposited in more recent sediments with placental mammals?

Alternatively, flood geologists suggest that fossil sequences are due to the separate submersion of different "pre-flood" biological communities. Studies of sedimentary deposits today demonstrate that land-dwelling dinosaurs

4. Snelling, "Order of Fossils."

lived in landscapes dominated by cycads, ginkgoes, and conifers (collectively called "gymnosperms"). In contrast, mammal fossils are commonly found in association with many varieties of flowering plants. Flood geologists suggest that the dinosaur-gymnosperm communities were laid down in their own sedimentary layers while the mammal-flowering plant communities were deposited in others. This sweepingly simple suggestion is not a realistic description of actual fossil deposits and it misrepresents how organisms are distributed in nature. A closer look shows that younger sediments, dominated by mammals and flowering plants, also contain representatives of gymnosperms and scattered types of even more primitive plants. On the other hand, the older dinosaurs-gymnosperm deposits *never* contain mammals or flowering plants precisely because the latter had not yet evolved.

Flood geology explanations entirely break down when we consider the locations of fossilized *plants*. Plants obviously lack the ability to move through floodwaters. What the fossil record shows is conspicuous *sequencing*: (1) the lowest sediments contain only primitive plants such as ferns and horsetails, (2) intermediate sediments are dominated by cone-bearing gymnosperms but also contain some ferns and horsetails, and finally, (3) recent sediments are filled with a huge variety of flowering plants, as well as representatives of all the earlier plant groups. These plant fossil patterns correspond to our comprehensive understanding of how plant types evolved over time.

Flood geology faces many other difficulties; particularly explaining how a worldwide flood only thousands of years ago (plus other natural processes since that time) could have formed all the geological features of the earth.[5]

Ark Passengers

There is also the issue of what animals Noah carried on the ark. The ark is described as 450 feet long, 75 feet wide, and 45 feet high (Gen 6:15). This is an enormous vessel (especially for ancient times), but it still would have limits in its capacity to carry eight humans, a variety of animals, and the food stores for these creatures. Whitcomb and Morris suggested a maximum capacity of 30,000 pairs of land animals. Biologists estimate there are over 4 million species of animals alive today and at least a quarter of them are land organisms (a large number are insects). Something doesn't add up. Also, there are those familiar representations of animals coming to board Noah's ark from all around the world, including alligators from North America, polar bears from the Arctic, giraffes from Africa, tigers from India,

5. Davidson and Wolgemuth, "Christian Geologists"; Isaak, "Problems"; Parkinson, "Questioning"; Senter, "Defeat."

and kangaroos from Australia. I would think that almost everyone who has pondered these depictions has wondered how all those creatures could possibly make it across oceans and deserts to the ark in the Near East.

Young-earth flood geologists attempt to solve the problem of limited space in ultimately unworkable fashions. First, they suggest that ark passengers may have been restricted to land vertebrates while all other types of life survived on floating mats of vegetation.[6] This would have severely reduced the diversity of animals other than vertebrates to repopulate the earth after the devastation of a global flood. Also, of the aquatic species left off the ark, almost all biologists would argue either most of the saltwater or most of the freshwater aquatic species would have been killed depending on the chemical composition of the universal floodwaters.

A second, and ultimately essential, proposal is that Noah took only one pair of each biblical "kind" and that every land vertebrate alive today descended from the original pairs by *biological evolution* over several thousand years. I've mentioned this proposal previously. For example, it has been suggested that one "cat" pair gave rise to all fossil and living members of the entire cat family, including domestic cats, lions, leopards, ocelots, cougars, bobcats, jaguars, cheetahs and tigers. Other diverse species groups were supposedly derived from single dog-like, horse-like, bear-like, and antelope-like pairs.[7] The advantage of this idea for flood geologists is that it requires relatively few parent pairs on the ark (one for each "kind") and this solves the logistical problem of limited space.

But how can the solution to the ark's limited size be a belief that biological evolution can proceed at an astounding pace well beyond anything ever understood or substantiated about evolutionary processes? The real irony is that most Christians who believe in a worldwide flood also believe that there has been relatively little biological evolution on earth. I am certain that many are unaware that the leaders of flood geology embrace remarkable amounts of rapid evolutionary change over a mere several thousand years.

Finally, the offered solution to worldwide animal migration to and from the ark is a mixture of speculations.[8] Flood geologists suggest there may have been a single pre-flood continental landmass that would have allowed free movement for animals to the ark. But there is absolutely no evidence that continents were joined only thousands of years ago. Post-flood migration has been explained as occurring across unidentified land bridges that appeared during times of lowered sea level or by massive uplifting and

---

6. Batten, *Creation Answers*, 181–88; Sarfati, "How Did All."

7. Batten, "Ligers and Wholphins?"

8. Batten, *Creation Answers*, 213–20; Taylor, "How Did Animals Spread."

subsiding of land formations. These land bridges would have had to form and disappear in the few thousand years since the flood event. There is little or no geological indication that most of these connections between continents ever existed. Further, the colossal geological events implied in these suggestions over such a very short timeframe should have been associated with incredible volcanism and global weather effects, effects that would have disrupted or destroyed normal functioning ecosystems.

Perhaps the most revealing evidence against flood geology is that none of its incredible claims are mentioned in the Bible. Does it not seem odd that there are no references in the narratives of Noah's flood or later Genesis chapters for (1) monumental atmospheric and geological events around the globe, (2) massive migrations of *unrecognizable* animal species from distant lands to Noah's location (i.e., most of the animals on the ark would have been previously unknown by Noah), or (3) remarkable evolutionary transformations in animal forms during the limited years of human history after Noah? It is far more reasonable to understand that, if Noah's ark was real, he gathered significant local animals onto his vessel and survived a large-scale flooding event in the region of the earth with which he and his family were familiar. There are many resources by Christian authors that provide good arguments for a local flood interpretation.[9]

Implications to Ponder

For many Christians, the mental picture of Noah's ark filled with animals from around the world is deeply seated, often starting with memorable childhood portrayals. It is always difficult to relinquish long-held imageries like this. However, an array of rational observations and alternative historical interpretations of Scripture obliges us to consider that the flood, although large and catastrophic in nature, was nonetheless localized in extent.

There is currently a strange disconnect within Christian churches; large numbers of adults understand that a global flood is not a well-supported idea and yet one of the most common Sunday school and children book lessons is a literal depiction of animals from across the earth entering Noah's ark. This is a good example of an established culture that deserves careful examination. Rather than accepting that this disconnect is innocuous or unimportant, I suggest that it should be proactively addressed by church leaders who understand the importance of staying relevant in an age that has obtained new insights about God's creation.

9. Hill, "Noachian Flood"; Young, "Biblical Flood."

I am certain that some believers will respond as if I am making a sac-rilegious suggestion. Nonetheless, I believe that inaccurate illustrations and play toys that show Noah's ark with recognizable animal species from far reaches of the earth either should be carefully clarified or they should be re-moved from Christian childhood education. Why do we raise our children to be confused in a real world of valid natural thought so that they will later be faced with either mistrusting scientific disciplines or questioning the foundations of their Christian faith? Why do adults let the significant theo-logical messages of sacred biblical text become confounded by the casual persistence of perspectives that directly conflict with revelations from the natural world? I suggest that the evangelical church can do more to be the dynamic, vibrant bride of the risen Lord. These issues are especially relevant if Christians hope to reach out to nonbelievers.

The crucial thing to remember in the story of the great flood and Noah's ark is its communication about God and his character. It reveals God's expression of grief and pain over human evil, wickedness, violence, and corruption (e.g. Gen 6:5–6, 11–12). It demonstrates the biblical prin-ciple of ultimate judgment passed by a holy and righteous God against evil. And, most importantly, it communicates God's mercy and care for those who turn to him to live in a personal relationship with him. Christians may unite on the significance of these assertions.

In Closing

I have attempted in the last three chapters to explain how many aspects of early Genesis can be interpreted in ways that coincide with current scien-tific understanding. I have done so because I believe the Bible is an entirely reliable communication from God to humanity and that the natural world is also the handiwork of the same Creator-God. The exercise is not about compromising Scripture to adapt to an increasingly decadent and immoral society. In a quite opposite and positive fashion, it is about Christians dis-covering how the unalterable foundations of faith revealed in the Bible har-monize with the uncovered scientific truths of God's creation. Evolutionary biologists should not have to solve theological issues for believers to validate their scientific work. Instead, I sincerely believe it is the job of the Christian church to work out a full theology of creation events, Adam and Eve, Origi-nal Sin, and the biblical flood in light of evolutionary understandings.

To conclude this section of the book, I recognize and am compelled to mention that Christians may attribute more significance than is warranted to aligning Genesis verses with modern scientific discoveries. We can be certain that the early pages of the Bible were never intended for scientific illumination and we will never fully understand the mindset of the peoples to which they were first written. The essential goal is to find a place that I believe God has for all of us where we can be comfortable with both the revelations of Scripture and scientific study. I am hopeful that my discussions have brought Christians to (or closer to) that position. We can expect that there will always be things we will not understand ("For we see in a mirror dimly . . . "—1 Cor 13:12), but we can also be content to continue on in faith, with thankfulness, as God has provided enough clarity to proceed along this path on which he has led us.

PART V

Maintaining Perspective

20

Christian Faith and Science

WE ARRIVE AT THE end of this book after traveling a path through an array of creation-evolution topics. These communications arise from my observations as a fully involved participant in both the conservative Christian church and the academic community of professional biologists, two worlds that are generally categorized as mutually exclusive. I have tremendous respect, admiration, and appreciation for individuals from both parts of my life. As a member of the evangelical church, I view the universe and our presence within it as the act of a sovereign, powerful, and loving Creator-God and I view the Bible as God's inspired message to humans, whom he created in his own image. I love my church family and I am deeply bonded with fellow Christians through a shared faith in what Jesus has done and continues to do for us. I am also a committed and enthusiastic biologist and I wholeheartedly accept and endorse the discoveries of modern science, including the fundamentals of evolutionary theory. I thoroughly enjoy my chosen profession and the fellow biologists who share my passion for the natural world.

The prevailing theme of this book has been that scientific discoveries and spiritual perspectives should not merely be allowed to coexist as separate, distinct spheres of knowledge or wisdom, but that science should be effectively integrated into a coherent all-inclusive worldview. I have challenged Christians and atheists, scientists and nonscientists to consider what this looks like and to think about what kinds of actions might lessen or abolish current detrimental divisions. The public conversation about creation and evolution has been dominated for far too long by individuals who do not represent the hearts and minds of most Americans. It is time to grab hold of the dialogue and direct it on a path toward healing.

I have explained how discoveries in the natural world solidly indicate a long evolutionary past. There is no real debate in the mainstream scientific community about this fundamental aspect of earth history. At the same time, the vast majority of Americans believe that God exists and most expect that this God has been involved in the creation of living things. My experience with a large diversity of people suggests that there is a strong desire for meaningful conceptualizations that bring these two understandings together. I have specifically described how we may embrace an enveloping reality that integrates the wonderful all-encompassing perspective of a biblical Creator-God with the valuable explanatory power in scientific descriptions of evolutionary biology.

Moving Forward without Confusion

There are a number of things that should happen for a balanced view of creation and evolution to grow and persist. The first is a rejection of propositions that either extrapolate beyond or invade into the distinctive nature of science. This includes both atheistic announcements based on scientific findings and God-centered explanations forced into scientific endeavors.

Atheism is ultimately a faith decision dependent on personal belief in the underlying assumption that nothing exists beyond properties of the natural world. But scientific discoveries cannot be used to prove this assumption, to conclude that God does or does not exist, or to answer other spiritual questions. I have argued that a very important starting point in creation-evolution discussions is acknowledging that all individuals develop their own understanding of our existence by accepting certain elements on a reasoned faith alone. In this way, an atheist view is fundamentally no different than a God-centered spiritual view; both are significantly derived from a nonmaterial sense about the validity of the position. Atheists have a right to their metaphysical choice, but it is inaccurate for it to be expressed as an obvious scientific judgment. Improper atheist communications demonstrate an unwillingness to acknowledge the spiritual choices of others. Effective dialogue and understanding can only occur when we respect one another's metaphysical decisions.

When atheism is expressed as an unavoidable conclusion from science, it also damages the integrity of the scientific discipline. I have explained that a truly workable integration of science and religion must include the foundation that science stays limited to studying natural parameters and processes. It is helpful to remember that only around 15% of Americans claim an atheistic viewpoint, so it remains a small minority voice. In contrast, deeply treasured religious faith has been an important aspect of intellectual life for most people throughout human history. The vast majority of us will continue to honor and embrace the nonscientific aspects of our existence while we simultaneously learn from new scientific advances.

A second extreme that should be avoided is strict young-earth creation that envisions that God created all living things over six 24-hour days some 10,000 years ago. I know it is impossible to make this recommendation without upsetting many evangelicals, but I am confident that an increasing number of believers will understand the compelling reasons for it. A young-earth viewpoint is based on one human-generated interpretation of Scripture that has not been an unwavering belief through the history of the Christian church. Although it can be extrapolated from a particular reading of Genesis passages, there is very little evidence in God's natural creation to support its validity. In contrast, evolutionary theory has developed over the last 150 years as one of the most thoroughly investigated, attacked, discussed, and refined sets of ideas in all of biology. It is the useful result of careful and sincere investigation of the natural world by both Christian and non-Christian scientists. These investigators have not been driven by biased presuppositions or an atheistic agenda. Despite claims to the contrary, the foundations of evolutionary biology are not being shaken by spontaneous creationist challenges. Evolutionary theory is also a deep and comprehensive set of ideas rather than a temporary conclusion within science that will be replaced by new information. To express the picture vividly (and with no intent to needlessly offend), scientists have no more reason to embrace young-earth creation than they do to return to an earth-centered view of our solar system.

This means that a strict young-earth creationist position does not integrate Christian faith with modern biology. It rests upon a dangerous presumption that a single interpretation of the Bible categorically falsifies sweeping agreement among practicing biologists. I have been enormously saddened by the effect this common stance has had on people both within and outside the church as conservative Christians isolate themselves and dig in against mainstream scientists. But I know there are many conservative believers who sense that there is something real to the evolutionary information that convinces so many biologists. Christians can proceed without fear, knowing that God's biblical and natural revelations will never conflict with one another. It is a Christian's task to incorporate legitimate scientific discoveries into a reasoned God-centered worldview, not the scientific community's job to conform to an unsupported biblical position. There are encouraging signs as increasing numbers of evangelical leaders are dialoguing with Christian scientists about the intersection of deep biblical faith and evolutionary biology.

Intelligent design propositions are different than young-earth creation because proponents typically combine a lot of modern science with their view of God's special creation actions. In fact, most of the originators and the most vocal ID advocates accept a very old age for the earth and openly acknowledge the significant role of evolutionary transformations. In this regard, intelligent design is potentially less hostile to goals of bringing together biblical Christian faith and modern science. On the other hand, ID arguments cause a different kind of crippling damage.

While suggesting that "complexity" and unexplained phenomena in nature are signs of God's special actions, intelligent design creationists are utilizing a "god-of-the-gaps" strategy that ultimately fails. First, ID arguments are not workable or helpful scientific hypotheses because, as God is inserted, they skip the fundamental criterion of explaining observations based on the working of natural mechanisms. Second, ID is an empty witness of God's activities because unexplained complexity or other gaps are not truly evidence of anything by themselves. Instead, we know that holes in current knowledge are historically filled in and that there is no intelligible way to designate that any current gap is unfillable. Even the theoretical possibility that new scientific explanations will arise to solve an ID creationist challenge shows that invoking God does not provide any clarity and can never really prove God's intervention or existence. Even further, why should Christians think we can demonstrate God only now within the intricate complexities of modern cellular biology when he has not already chosen to unequivocally display his intervening hand in our vast knowledge of whole organisms? We have ample evidence that God does not interact with his natural creation in the way ID advocates propose. We must conclude that intelligent design inappropriately attacks and redefines the fundamental character of scientific endeavors and it is ineffective in demonstrating God in the natural world.

I know the arguments I have presented against young-earth creation and intelligent design can be difficult to accept for Christian individuals who sincerely want to see more definitive statements about God's actions within natural processes. Even so, a strong desire for such a thing does not make these strategies valid or effective. But God is not absent!

Christians may proclaim that God is the sovereign Creator of the universe and that he is the author of biological evolution. He most certainly is the primary cause of all living things as he created and maintains the mechanisms by which evolution operates. Even further, the Bible itself indicates that God's creative work is enacted *through* these natural processes that he continually sustains. Our understanding of divine action must allow for a God who is outside his creation, and who transcends both space and time, rather than depending only on the limited cause-and-effect descriptions that humans are familiar with on earth. We cannot know the precise nature of God's influence in evolutionary change because he has not left obvious signs of his involvement in ways that are distinguishable from the results of natural processes. Christians can be comfortable with this fact in the same way that we accept that God in his wisdom has not otherwise revealed himself to all people of the earth today in a blatant and undeniable manner. Christians may embrace the understanding that science only uses naturalistic explanations and does not address the existence or actions of God. Even so, God is readily apparent in his natural creation by a simple nonscientific ascension to the obvious: the heavens declare the glory of God! No passionate arguments by atheistic scientists will ever negate our common experiences of beauty, awe, and wonder or our sense, our knowledge, that there is something more, a spiritual reality and a God that is behind all aspects of our existence.

An Effective Posture for Scientists

I challenge fellow scientists to examine where and how they have contributed to the creation-evolution divide that exists today. Within the pages of this book, I have strongly defended the validity and explanatory importance of evolutionary theory and the purity of naturalistic explanations within scientific disciplines. The crucial question that remains is how these fundamental understandings are integrated with the spiritual perspectives of individuals.

An antireligious culture currently exists within the modern scientific community. This ranges from unspoken expectations that religious views are not discussed, to subtle or overt forms of religious suppression and condescension, to outward pronouncements that science has proved an atheistic worldview. This culture is not simply explained by antireligious sentiments among many of the participants since a significant number of scientists acknowledge spiritual experiences and religious affiliations. A major influence is undoubtedly the prevailing but improper assumption that science and religion are distinct, mutually exclusive spheres of wisdom. This approach is not, however, the understanding of the majority of Americans who inherently know the two areas of knowledge cannot be isolated from each other.

Scientists can proceed with an increased awareness that they operate within an American culture dominated by individuals with religious,

especially Christian, worldviews. Scientists should understand that the naturalistic methodology of our discipline, and the non-spiritual tenor of the
scientific community, projects an unwelcoming communication for those
who strive to incorporate their religious faith into their everyday lives. Even
though there are very valid scientific reasons for rejecting creation science
and intelligent design from scientific curricula, it should be apparent that
many people will misunderstand these judgments as the ongoing exclusion
of God and spirituality from public dialogue. If the projected face of the
scientific establishment is atheistic, there is little reason to expect Christian
individuals to accept scientific discoveries that pose difficulties for their
current religious viewpoints. This means that the reception of many exciting
scientific discoveries will often depend on how well scientists act in accordance with this recognition. In sum, if the scientific community does not alter its atheistic posture, it significantly contributes to the creation-evolution
divide and continued controversy.

Scientists, in general, can more outwardly embrace the legitimate
value of spiritual perspectives and show respect for and acceptance of
varied belief systems. Fellow scientists should reject exclusionary atheistic
pronouncements by colleagues. This is a call for individuals with atheist and
agnostic views, not just religious scientists, to speak up and alter the cultural
tenor within science institutions, particularly across academia. I also suggest that all scientists should consider where they may be purposefully or
inadvertently communicating atheism or downgrading spiritual views.

The dynamic I am addressing is most notably played out in the arena
of science education. Biology instructors can acknowledge, with sensitivity,
that the findings of science will have to be integrated into each individual's
faith-based worldview. This is not only doable, it is right to embrace and promote spiritual/metaphysical awareness in this fashion. I have suggested that
science educators should receive at least some kind of appropriate training
in the philosophy of science and science-spirituality issues. They can then
include a short but careful presentation on the interrelationship of science
and religion in introductory science classes. Instructors absolutely should
not include spontaneous creationist propositions in biology curricula in an
attempt to be "fair" because this distorts the very nature of science. Even
so, science educators must not work so hard to protect the empirical facts
of science that they forget or are insensitive to the religious questions or
emotional responses those facts generate.

Over my years of college instruction, I have had numerous Christian
students enter my biology courses with an accepted young-earth creation
viewpoint. Some expressed fear or were guarded about what I would say
about evolution. Some gave me creation science and/or intelligent design

materials to demonstrate different views. But all these students heard my introductory comments within class that evolution and deep religious faith are compatible. I expressed a gentle understanding that evolution and other scientific discoveries may challenge religious beliefs and raise significant, even disturbing, questions. I demonstrated that I was sensitive to the experiences of students who felt uncomfortable and I was willing to discuss challenges with them if they wanted. In turn, I asked all students to be open to naturalistic evolutionary information and to examine it at its face value. Invariably, I was thanked by both Christian and non-Christian students for my direct and candid presentation.

I cannot report that all young-earth creationist students altered their original perceptions, but I can happily state that the majority slowly became convinced by the evolutionary story scattered across the biological landscape. Important for Christians to know, the faith of these individuals was not shattered by evolutionary information. Instead, they were equipped with a dynamic worldview that includes both evolution and Christian faith, that is, evolutionary creation. The point for science educators, of course, is that these students were reached with sensitive instruction aimed at a holistic understanding of their experiences and perceptions.

A Vibrant Christian Message

Christians are united in their recognition that the Bible is God's revelation to humans through the inspired authorship of its writers. It contains ultimate truths about the world and, most importantly, our relationship to the Creator. It significantly communicates that God is the foundation behind our existence and the Creator and Sustainer of the natural world. Believers may also unite in the strong Christian tradition that God's natural creation is an additional revelation to us about this Creator-God. Given the substantial evidence for a very old universe and a long history of biological evolution, we are drawn with good reason to interpret the Bible in ways that are compatible with these concepts. This endeavor should be pursued, not to fit God into a specific worldview, but to know and appreciate an all-encompassing view of the God we worship based on all the ways he has revealed himself to us. In evolutionary creation, the depth and richness of God are most fully revealed, like hearing the beautiful interplay of two hands at a piano instead of one.

It is helpful for Christians to remember and follow prominent theologians from the past, such as Augustine, Aquinas, and Calvin, who understood that knowledge about God was an ongoing enlightenment. Their walk with Christ included celebrating new discoveries in the natural world and integrating them with the words of Scripture. Although there are certainly foundational doctrines of Christian faith that cannot be compromised, these church leaders demonstrate that real Christian orthodoxy is to work with all information in open and honest intellectual pursuit, not to hold fast to established perceptions. We have been instructed by historical events, like the debate over a sun-centered solar system, that Scriptural interpretations about the natural world by themselves can be inaccurate and misleading.

Christians can proceed by listening carefully and excitedly to hear and understand what scientists are revealing in their work. We must be compelled to reexamine interpretations of the Bible that conflict with any valid discoveries. In this way, believers should honestly wrestle with the findings of evolutionary science and investigate how evolutionary creation provides our most complete view of the God of the universe. This is not weakness or defeat; it is healthy intellectual engagement by a strong, vibrant, and relevant Christian church.

Evolutionary creation should be embraced not only for its consistency with modern science but for the sake of those who have not seriously considered the claims of Jesus. America is full of individuals who are convinced of the general reality of biological evolution. Most have also heard that young-earth creation or some other spontaneous creationist derivation is the unquestioned teaching of the Bible. These people are not inherently closed to spiritual discussions or the possibility of a living God who was and is at work, but they are driven away from the Bible and Christ by the pervasive communication that Scripture teaches something that the scientific evidence shows is false. This negative impact is particularly strong in the scientific community where its members have been accused of dishonesty and deception. Augustine strongly cautioned against this error way back in the 5th century and Aquinas and others have repeated his message through the history of the church. Augustine stated:

> Now it is an unseemly and mischievous thing, and greatly to be avoided, that a Christian man speaking on matters, as if according to authority of Christian Scripture, should talk so foolishly that an unbeliever on hearing him, and observing the extravagance of his error, should hardly be able to refrain from laughing. And the great mischief is not so much that the man is laughed at for his errors, but that our biblical authors are believed, by people without the Church, to have taught such things, and are so condemned as unlearned, and cast aside, to the great loss of those for whose salvation we are so much concerned.
>
> For when they find one belonging to the Christian body falling into error on a subject with which they themselves are

thoroughly conversant, and when they see him moreover en-
forcing his groundless opinion by the authority of our Sacred
Books, how are they likely to put trust in these Books about
the resurrection of the dead, and the hope of eternal life, and
the kingdom of heaven, having already come to regard them as
fallacious about those things they had themselves learned from
observation, or from unquestionable evidence?[1]

The very sad reality is that numerous non-Christians have indeed
"laughed" at the evangelical church. Believers should be very concerned
that the messages of the Bible have been cast aside precisely as described in
Augustine's warning.

The real job for Christians is to explain how Jesus remains relevant
no matter what the scientific community ascertains. We should focus on
demonstrating the need for God even as we live within the context of a
scientific world. As Christians, we have one overriding purpose: to praise
and glorify Jesus Christ and to offer the message of life through Him to
those who might hear. The key communication of creation is that God is
all-powerful and sovereign over all things. The important question is not
what methods God used to create us, but whether he created us at all!

In the previous section, I reported how conservative Christian students in my college courses changed their views when exposed to sensitive teaching of evolutionary theory. The mirrored experience has been
non-Christian students who drop by or are brought to my office to discuss
evolution and biblical faith. The typical scenario is an individual who already acknowledges evolutionary concepts, but has been led to inquire into
Christianity. The bottom line has often been, "Can I accept both evolution
and the claims of Jesus?" It has been a joy for me to explain whatever the
person wanted to hear about evolutionary creation and to encourage and
direct him or her in their continued pursuit of Christian truths.

Thoughtful Transitions

The Christian church can contribute to meaningful healing in the creation-
evolution divide by altering entrenched behaviors that alienate spiritual
seekers and other Christians who acknowledge an evolutionary past. One
important domain is the music we sing together. A large number of traditional hymns and contemporary songs contain references to God bringing all things into being and to the incomprehensible magnitude and

1. Augustine, *Literal Meaning*, Book I, Chapter 19, in Molley, *Geology and Revelation*, 297.

overwhelming splendor in nature. Most of these rightly bring us to a place of praise and worship. However, there is no reason to include songs (no matter how loved they may be) that contain disputable descriptions of God creating things in an instant or fashioning them in the time frame of days.[2]

Another notable area for examination is the education of our children.[3] Christians not only want our youth to acquire their own meaningful understanding of Christ-centered faith but also to grow to engage effectively with the world in which they live. Many current approaches invariably lead to crises of faith when young adults encounter strong disharmony between science and their Christian schooling. In one simple illustration, children should not be taught about a worldwide flood as if it was a real event with pictures and toys depicting Noah's ark filled with animals from all around the earth. Other topics that produce unneeded conflict can also be reevaluated. Parents should be mindful to guide their children to a healthy understanding that scientific studies of the natural world and the Bible both have it right!

Pastors who are only seminary-trained in young-earth creation perspectives must be challenged to construct messages that are broadly sensitive. Rather than falling into a comfort zone of expected evangelical references, they should consider how unexamined comments are often unwelcoming to Christians and seekers who have very good reasons to understand earth history differently. Although proposals like these may be unconventional, and difficult notions for Christians accustomed to special creation perceptions, our focus should not be on the comfort of tradition-bound members of the Christian family. Instead, it should be on individuals outside the church who are drawn to the light of the gospel.

Finally, the Christian church can do more in a positive direction to celebrate scientific discoveries. Although the methods of science exclude God from its explanations, the opposite is not at all true, that Christians should exclude scientific knowledge from our understanding and worship of God. The discovery of a new genetic mechanism, a new fossil tree, or a new comet should excite us as fascinating aspects of the world God created. Part of the transition to an evolutionary creation worldview is an understanding that God may be worshipped freely and wholeheartedly *because* of the latest scientific advances, not in spite of them. Fortunately, there are increasing numbers of resources for pastors and laypeople to help bring the exciting revelations of science into the everyday world of the church.[4]

2. As an example, I like to use the beloved hymn "At The Name of Jesus" when I lead worship at my church, but I skip a verse that begins: "At His voice creation sprang at once to sight . . . "

3. Falk, "Saving the Children."

4. BioLogos Resources; Cosmos: Refaithing Science; Huyser-Honig, "Science and

These include sermons with deep scientific references, science updates by Christian authors, Christian science curricula, and ways to include scientific discoveries in worship.

I encourage Christians who would like to participate in effecting a transition to openly communicate that real reconciliation between mainstream science and Christian faith is possible and desirable. Ask friends, church leaders, teachers, and pastors to investigate and understand evolutionary creation. Express disappointment when individuals and organizations assume that spontaneous creationists positions are the only formulations compatible with Christian faith. Question why so many lectures, seminars, bible studies, and debates are based on the false dichotomy that spontaneous creationists are Christians and evolutionists are atheists. Hold out for discussions that bring together modern scientific findings and Christianity and ask Christian organizations and media to promote these conversations. Suggest to Christian bookstores that they carry evolutionary creation offerings alongside the spontaneous creation books everyone expects to see.

Based on my own experiences, by doing any of these things, individuals should realistically prepare for pressure or ridicule from some Christian brothers and sisters who would not allow believers to accept such things. But if you trust as I do that God would want something different from his church than the current damaging controversy, it is possible to alter the spontaneous creationists' perceptions that currently dominate evangelical thought. It is my hope and prayer that more Christians can join the ranks of those who see God's creative power and biological evolution as harmonious. It is by this transformation that many more people will consider the life-changing claims of Jesus.

Closing Thoughts

I conclude by returning once more to the sense of awe, inspiration, and beauty that is commonly experienced within natural settings. There is nothing quite like the feelings we have if we notice the delicate pattern of veins on a leaf, behold the brilliant colors of a peacock tail, or marvel at the overwhelming number of stars in a night sky. It is fitting to end a book on bridging differences by celebrating this universal aspect of our human condition. Regardless of an individual's spiritual perspective or scientific background, we all seem to understand that we are part of a natural world of infinite wonder. This is a wonderful phenomenon.

Faith"; The Ministry Theorum; Scientists in Congregations; Wiseman, "Instrument of Worship."

From this common ground of shared experience, each of us is allowed to choose how these sensations are explained or to what they are attributed. Respect for this choice will be a primary determinant in how well we may reach across divisions between Christian and scientific communities. Some believe inspirational experiences are aspects of our neural circuitry programmed into perceptions as a product of evolutionary history. Others have spiritual explanations. For Christians, beauty and majesty in nature are an indication that the natural world is a creation of an active, living God.

I found a place of relief, joy, and certainty when I became persuaded that there is something more than natural mechanisms at work and conceded that God is behind it all. The Bible's message is that God's presence and power in nature are evident to us without any reference at all to scientific information. A spiritual decision always involves the heart, as well as the intellect, and requires a pursuit of truth based on faith in something supernatural. I was once a scientist who saw the world as devoid of spiritual influence, but the magnificent and awe-inspiring aspects of nature were a key invitation for me to consider the God of the Bible and the claims of Jesus. The invitation is expressed like this in Isaiah 42:5–6: "This is what God the LORD says—he who created the heavens and stretched them out, who spread out the earth and all that comes out of it, gives breath to its people, and life to those who walk on it: 'I, the LORD, have called you in righteousness; I will take hold of your hand.'"

Bibliography

All online resources last accessed during April and May 2014.

Ahlberg, Per E., and Jennifer A. Clack. "A Firm Step From Water to Land." *Nature* 440 (2006) 747–49.

Alexander, Denis R. *Creation or Evolution: Do We Have To Choose?* Grand Rapids: Monarch, 2008.

Alter, Jonathan. "Monkey See, Monkey Do." *Newsweek*, August 15, 2005, 27.

American Association for the Advancement of Science. *A Study Guide for The Evolution Dialogues: Science, Christianity and the Quest for Understanding.* Washington DC: AAAS, 2006.

———. *Dialogue on Science, Ethics and Religion.* http://www.aaas.org/DoSER

Aquinas, Thomas. *Summa Theologica*, ca. 1274. Translated by Fathers of the English Dominican Province, 1920. http://www.newadvent.org/summa/*Summa*

Arthur, Wallace. "The Emerging Conceptual Framework of Evolutionary Developmental Biology." *Nature* 415 (2002) 757–64.

Ashcraft, Chris. "Evolution; God's Greatest Creation." Northwest Creation Network. http://nwcreation.net/evolution_creation.html

Astin, Alexander W., et al. *Cultivating The Spirit: How College Can Enhance Students' Inner Lives.* San Francisco: Jossey-Bass, 2010.

Augustine. *The City of God.* Translated by Marcus Dods. 1887. http://www.newadvent.org/fathers/1201.htm

———. *The Literal Meaning of Genesis.* Translated by John H. Taylor. New York: Newman, 1982.

Ayala, Francisco J. *Darwin and Intelligent Design.* Minneapolis: Fortress, 2006.

Barbour, Ian. *Religion and Science.* San Francisco: HarperCollins, 1997.

Barr, James. 1992. *The Garden of Eden and the Hope of Immortality.* Minneapolis: Fortress 1992.

Bartholomew, David J. *God, Chance and Purpose: Can God Have It Both Ways?* Cambridge University Press, 2008.

Bartz, Paul A. "The Religion of Evolution." *Bible Science Newsletter.* Foley, MN: Creation Moments, 1988.

Batten, Don, ed. *The Creation Answers Book.* Powder Springs, GA: Creation Book, 2007

Batten, Don. "Ligers and Wholphins? What Next?" *Creation* 22:3 (2000) 28–33.

Beale, Gregory K. "Eden, the Temple, and the Church's Mission in the New Creation." *Journal of the Evangelical Theological Society* 48 (2005) 5–31.

———. *The Temple and the Church's Mission: A Biblical Theology of the Dwelling Place of God*. Downers Grove, IL: InterVarsity, 2004.

Behe, Michael J. *Darwin's Black Box: The Biochemical Challenge to Evolution*. New York: Free Press, 1996.

———. *The Edge of Evolution: The Search for the Limits of Darwinism*. New York: Free Press, 2007.

Benton, Michael J. *Vertebrate Paleontology*. 3rd ed. Hoboken, NJ: Wiley, 2004.

BioLogos Foundation. "Your Stories, Dr. Stephen L. Pelton." http://biologos.org/about/stories/dr-stephen-l-pelton

———. *Statement of Participants at the BioLogos Workshop, "In Search of a Theology of Celebration."* November 12, 2009. http://biologos.org/uploads/projects/Workshop_statement.pdf

———. Resources. http://biologos.org/resources

Blocher, Henri. *In the Beginning: The Opening Chapters of Genesis*. Downers Grove, IL: InterVarsity, 1984.

Boisserie, Jean-Renaud, et al. "The Position of Hippopotamidae Within Cetartiodactyla." *Proceedings of the National Academy of Science* 102 (2005) 1537–41.

Buller, David. "Video Testimony." BioLogos, October 2012. http://biologos.org/newsletter/2012/10/video-testimony.html

Calvin, John. *Commentary on Genesis*, 1554. Translated by John King. 1578. London: Banner of Truth Trust, 1975. http://www.iclnet.org/pub/resources/text/ipb-e/epl-cvgenesis.html

———. *Institutes of the Christian Religion*, 1559. Translated by Henry Beveridge, 1845. http://www.reformed.org/master/index.html?mainframe=/books/institutes/

Camp, Ashby L. "The Overselling of Whale Evolution." *Creation Matters*, May/June 1998.

Campbell, Neil A., et al. *Biology: Concepts and Connections*. 2nd ed. Menlo Park, CA: Benjamin-Cummings, 1997.

———. *Biology: Concepts and Connections*. 4th ed. San Francisco: Benjamin-Cummings, 2003.

Carroll, Robert L. *Vertebrate Paleontology and Evolution*. New York: W.H. Freeman, 1988.

Carroll, Sean B. *Endless Forms Most Beautiful: The New Science of Evo Devo*. New York: Norton, 2006.

Carroll, William E. "Creation, Evolution and Thomas Aquinas." *Revue des Questions Scientifiques* 171 (2000) 319–47. http://www.catholiceducation.org/articles/sc0035.html

Catchpoole, David, and Carl Wieland. "Speedy Species Surprise." *Creation* 23:2 (2001) 13–15.

Cedars-Sinai Medical Center. ""Defective" But Beneficial Gene May Bring About Novel Ways To Clear Arterial Plaque Buildup." *Science Daily*, February 17, 2000. http://www.sciencedaily.com/releases/2000/02/000216082501.htm

Center for the Renewal of Science and Culture. *The Wedge*. Seattle: Discovery Institute, 1999. http://www.antievolution.org/features/wedge.pdf

Chandler, Matt. *The Explicit Gospel*. Wheaton, IL: Crossway, 2012.

Chandler, Vicki L. "Paramutation: From Maize to Mouse." *Cell* 128 (2007) 641–45.

Chapell, Bryan. *'98-'99 President's Goals and Report*. Covenant Theological Seminary, 1998. http://www.reformed.org/creation/index.html

Chiappe, Luis M. *Glorified Dinosaurs*. New York: Wiley-Liss, 2007.

Chimpanzee Sequencing and Analysis Consortium. "Initial Sequence of the Chimpanzee Genome and Comparison With the Human Genome." *Nature* 437 (2005) 69-87.

Clack, Jennifer A. "Getting A Leg Up On Land." *Scientific American*, December 2005, 100–07.

Coates, M.L., and Jennifer A. Clack. "Fish-like Gills and Breathing In the Earliest Known Tetrapod." *Nature* 352 (1991) 234–36.

Collins, C. John. "Adam and Eve as Historical People, and Why It Matters." *Perspectives on Science and Christian Faith* 62 (2010) 160–61.

———. *Did Adam and Eve Really Exist?: Who They Were and Why You Should Care*. Wheaton, IL: Crossway, 2011.

———. *Genesis 1-4: A Linguistic, Literary and Theological Commentary*. Phillipsburg, PA: Presbyterian and Reformed, 2006.

Collins, Francis S. *Belief: Readings on the Reason for Faith*. New York: HarperOne, 2010.

———. *The Language of God: A Scientist Presents Evidence for Belief*. New York: Free Press, 2006.

Collins, Robin. "Evolution and Original Sin." In *Perspectives on an Evolving Creation*, edited by Keith B. Miller, 469–501. Grand Rapids: Eerdmans, 2003.

Conway-Morris, Simon. "Darwin was Right. Up to a Point." *The Guardian*, February 12, 2009. http://www.guardian.co.uk/global/2009/feb/12/simon-conway-morris-darwin

———. *Life's Solution: Inevitable Humans in a Lonely Universe*. New York: Cambridge: Cambridge University Press, 2003.

Copley, Shelley D. "Evolution of a Metabolic Pathway for Degradation of a Toxic Xenobiotic: the Patchwork Approach." *Trends in Biochemical Science* 25 (2000) 261–65.

Cosmos: Refaithing Science. Regent College. http://cosmos.regent-college.edu/

Culotta, Elizabeth. "Chimp Genome Catalogs Differences With Humans." *Science* 309 (2005) 1468–69.

D'Souza, Dinesh. *What's So Great About Christianity*. Washington DC: Regnery, 2008.

Daeschler, Edward B., et al. "A Devonian Tetrapod-like Fish and the Evolution of the Tetrapod Body Plan." *Nature* 440 (2006) 757–63.

Darwin, Charles. *On The Origin of Species*. London: John Murray, 1860. http://darwin-online.org.uk/

Davidson, Gregg, and Ken Wolgemuth. "Christian Geologists on Noah's Flood: Biblical and Scientific Shortcomings of Flood Geology." BioLogos, July 2010. http://biologos.org/uploads/projects/davidson_wolgemuth_scholarly_essay.pdf

Davis, Percival, and Dean Kenyon. *Of Pandas and People*. 2nd ed. Mesquite, TX: Haughton, 1993.

Dawkins, Richard. *The Blind Watchmaker*. New York: Norton, 1986.

———. *The God Delusion*. Boston: Houghton Mifflin, 2006.

———. *The Magic of Reality: How We Know What's Really True*. New York: Free Press, 2012.

Day, Allen J. "Adam, Anthropology and the Genesis Record – Taking Genesis Seriously in the Light of Contemporary Science." *Science and Christian Belief* 10 (1998) 115–43.

Del Hoyo, Josep, Andrew Elliott and David Christie, eds. *Handbook of the Birds of the World, Vols. 8–16*. Barcelona, Spain: Lynx Edicions, 2003–2011.

Dembski, William A. *The Design of Life: Discovering Signs of Intelligence in Biological Systems*. Richardson, TX: Foundation for Thought and Ethics, 2007.

———. *Intelligent Design: The Bridge Between Science and Theology*. Downers Grove, IL: InterVarsity, 1999.

———. "Introduction: Mere Creation." In *Mere Creation: Science, Faith and Intelligent Design*, edited by William A. Dembski, 13–30. Downers Grove, IL: InterVarsity, 1998.

———. *No Free Lunch: Why Specified Complexity Cannot Be Purchased without Intelligence*. Lanham, MD: Rowman and Littlefield, 2001.

———. "Southern Baptist Voices: Is Darwinism Theologically Neutral?" BioLogos, April 30, 2012. http://biologos.org/blog/southern-baptist-voices-is-darwinism-theologically-neutral

Dennett, Daniel. *Breaking the Spell: Religion as a Natural Phenomenon*. New York: Penguin, 2007.

Devor, Eric J., et al. "Serine Hydroxymethyltransferase Pseudogene, SHMT-ps1: A Unique Genetic Marker of the Order Primates." *Journal of Experimental Zoology* 282 (1998) 150-56.

Easterbrook Philippa J. "Long-Term Non-Progression in HIV Infection: Definitions and Epidemiological Issues." *Journal of Infection* 38 (1999) 71–73.

Elphick, Michael. "Baraminology" FalseOrigins, November 11, 2011. http://www.falseorigins.org/baraminology.html

ENCODE Project Consortium. "An Integrated Encyclopedia of DNA Elements in the Human Genome." *Nature* 489 (2012) 57-74.

Enns, Peter. *The Evolution of Adam: What the Bible Does and Doesn't Say About Human Origins*. Grand Rapids: Brazos, 2012.

———. *Inspiration and Incarnation: Evangelicals and the Problem of the Old Testament*. Grand Rapids: Baker Academic, 2005.

Expelled. Directed by Benjamin Stein. Universal City, CA: Vivendi Visual Entertainment, 2008.

Falk, Darrel R. *Coming to Peace With Science: Bridging the Worlds Between Faith and Biology*. Downers Grove, IL: InterVarsity, 2004.

———. "Does Genetics Point to a Single Primal Pair?" BioLogos, April 5, 2010. http://biologos.org/blog/does-genetics-point-to-a-single-primal-couple

———. "Joanna's Story." BioLogos, December 23, 2010. http://biologos.org/blog/gloriain-excelsis-deo

———. "One Hundred Fifty Years . . . and Counting." BioLogos, November 30, 2009. http://biologos.org/blog/one-hundred-and-fifty-years-and-counting

———. "Saving the Children." BioLogos, October 5, 2009. http://biologos.org/blog/saving-the-children

Fedoroff, Nina V. "Transposable Elements, Epigenetics, and Genome Evolution." *Science* 338 (2012) 758–67.

Forrest, Barbara, and Paul R. Gross. *Creationism's Trojan Horse: The Wedge of Intelligent Design*. Oxford: Oxford University Press, 2004.

Friar, Wayne. "Baraminology — Classification of Created Organisms." *Creation Research Society Quarterly* 37:2 (2000) 82–91.

Gatesy, John. "More Support For a Cetacea/Hippopotamidae Clade: the Blood Clotting Protein Gene g-fibrinogen." *Molecular Biology and Evolution* 14 (1997) 537-43.

Gatesy, John, et al. "Evidence from Milk Casein Genes That Cetaceans Are Close Relatives of Hippopotamid Artiodactyls." *Molecular Biology and Evolution* 13 (1996) 954-63.

―――. "Stability of Cladistic Relationships Between Cetacea and Higher-Level Artiodactyl Taxa." *Systematic Biology* 48 (1999) 6–20.

Giberson, Karl. *Saving Darwin: How To Be a Christian and Believe in Evolution.* New York: HarperOne, 2008.

Giberson, Karl W., and Francis S. Collins. *The Language of Science and Faith: Straight Answers to Genuine Questions.* Downers Grove, IL: InterVarsity, 2011.

Gingerich, Philip D. "Evolution of Whales from Land to Sea." *Proceedings of the American Philosophical Society* 156 (2012) 309–23.

Gingerich, Philip D., and Mark D. Uhen. "*Ancalecetus simonsi*, A New Dorudontine Archaeocete (Mammalia, Cetacea) from the Early Late Eocene of Wadi Hitan, Egypt." *Contributions from the Museum of Paleontology, University of Michigan* 29 (1996) 359–401.

Gingerich, Philip D., et al. "Hind Limbs of Eocene *Basilosaurus*: Evidence of Feet in Whales." *Science* 249 (1990) 154–57.

―――. "New Protocetid Whale from the Middle Eocene of Pakistan: Birth on Land, Precocial Development, and Sexual Dimorphism." *PLoS ONE* 4 (2009): e4366.doi: 10.1371/journal.pone.0004366.

―――. "New Whale from the Eocene of Pakistan and the Origin of Cetacean Swimming." *Nature* 368 (1994) 844–47.

―――. "Origin of Whales From Early Artiodactyls: Hands and Feet of Eocene Protocetidae from Pakistan." *Science* 293 (2001) 2239–42.

Gishlick, Alan D. "Baraminology." *Reports of the National Center for Science Education* 26:4 (2006) 17–21.

―――. "Icons of Evolution?: Why Much of What Jonathan Wells Writes About Evolution Is Wrong." National Center for Science Education, October 19, 2008. http://ncse.com/creationism/analysis/icons-evolution

Gould, Stephen J. "Nonoverlapping magisteria." *Natural History*, March 1997, 16–22.

―――. *The Structure of Evolutionary Theory.* Cambridge, MA: Belknap, 2002.

Graham, Linda E. *The Origin of Land Plants.* New York: Wiley, 1993.

Graham, Linda E., et al. "Phylogenetic Connections Between the 'Green Algae' and 'Bryophytes.'" *Advances in Bryology* 4 (1991) 213–44.

Gray, Terry M. "Biochemistry and Evolution." In *Perspectives On An Evolving Creation*, edited by Keith B. Miller, 256–87. Grand Rapids: Eerdmans, 2003.

Haarsma, Deborah, and Loren Haarsma. *Origins: Christian Perspectives on Creation, Evolution and Intelligent Design.* 2nd ed. Grand Rapids: Faith Alive Christian Resources, 2011.

Ham, Ken. "The Evil Fruit of Evolutionary Thinking." In *The Lie: Evolution*, 137–54. Green Forest, AR: Master, 2012.

―――. "I Know Nothing." Answers in Genesis, February 22, 2002. http://www.answersingenesis.org/articles/au/i-know-nothing

―――. "Searching for the Magic Bullet." *Creation* 25:2 (2003) 34-37.

————. "What's the Best "Proof of Creation?" *The New Answers Book* 2, Chapter 2. Green Forest, AR: Master, 2008.

————. "A Young Earth - It's Not the Issue!" Answers in Genesis, January 23, 1998 https://answersingenesis.org/why-does-creation-matter/a-young-earth-its-not-the-issue/

Harlow, Daniel C. "After Adam: Reading Genesis in an Age of Evolutionary Science." *Perspectives on Science and Christian Faith* 62 (2010) 179–95.

Harris, Sam. *The End of Faith: Religion, Terror, and the Future of Reason*. New York: Norton, 2005.

————. *The Moral Landscape: How Science Can Determine Human Values*. New York: Free Press, 2010.

Hastings, Ronnie J. "The Rise and Fall of the Paluxy Mantracks." *Perspectives on Science and Christian Faith* 40 (1988) 144–54

Haught, John F. *God and the New Atheism: A Critical Response to Dawkins, Harris, and Hitchens*. Louisville: Westminster John Knox, 2007.

Hayward, Alan. *Creation and Evolution*. Eugene, OR: Wipf & Stock, 1985. Quoted in Stephen J. Gould. "Hooking Leviathan By Its Past." *Natural History*, May 1994, 8–15.

Higher Education Research Institute. *Spirituality and the Professoriate: A National Survey of Faculty Beliefs, Attitudes and Behaviors*. UCLA, 2005.

Hill, Carol A. "The Noachian Flood: Universal or Local?" *Perspectives on Science and Christian Faith* 54 (2002) 170-83.

Hitchens, Christopher. *God Is Not Great: How Religion Poisons Everything*. New York: Twelve, 2007.

Hitchens, Peter. *The Rage Against God: How Atheism Led Me to Faith*. Grand Rapids: Zondervan, 2010.

Hodge, Archibald A., and Benjamin B. Warfield. "Inspiration." *The Presbyterian Review* 6 (1881) 225-60.

Hoggard, Michael W. "Man Walked With Dinosaurs," YouTube, December 15, 2010. http://www.youtube.com/watch?v=dtqtgUsyrj8

Hunt, Kathleen. "Transitional Vertebrate Fossils FAQ." TalkOrigins Archive, 1997. http://www.talkorigins.org/faqs/faq-transitional.html

Hurd, James P. "Hominids in the Garden?" In *Perspectives on an Evolving Creation*, edited by Keith B. Miller, 208–33. Grand Rapids: Eerdmans, 2003.

Huyser-Honig, Joan. "Science and Faith in Harmony: Positive Ways to Include Science in Worship." Calvin Institute of Christian Worship, November 4, 2009. http://worship.calvin.edu/resources/resource-library/science-and-faith-in-harmony-positive-ways-to-include-science-in-worship/

Irwin, Darren E., et al. "Speciation in a Ring." *Nature* 409 (2001) 333–37

Isaak, Mark. *The Counter-Creationism Handbook*. Berkeley: University of California Press, 2007.

————. "Problems with a Global Flood." TalkOrigins Archive, November 16, 1998. http://www.talkorigins.org/faqs/faq-noahs-ark.html

Issac, Randy. "Science and The Question of God." BioLogos, September 2010. http://biologos.org/resources/essay/science-and-the-question-of-god

Jablonka, Eva, and Gal Raz. "Transgenerational Epigenetic Inheritance: Prevalence, Mechanisms, and Implications for the Study of Heredity and Evolution." *Quarterly Review of Biology* 84 (2009) 131–76.

Jeanson, Nathaniel T. "Presuppositional Research and Definition of Kind." *Acts & Facts* 39:10 (2010) 8.

Johnson, Philip E. *Darwin on Trial*. Downers Grove, IL: InterVarsity, 1993.

———. "Shouting 'Heresy' In the Temple of Darwin." *Christianity Today*, October 24, 1994, 22–26.

Joiner, William. "Creationist Mindblocks to Whale Evolution." 1999. http://www.angelfire.com/fl/direpuppy/mindblocks.html

Kaessmann, Henrik. "Origins, Evolution, and Phenotypic Impact of New Genes." *Genome Research* 20 (2010) 1313–26.

Kardong, Kenneth V. *Vertebrates: Comparative Anatomy, Function, Evolution*. 6th ed. New York: McGraw Hill, 2011.

Karol, Kenneth G., et al. "The Closest Living Relatives of Land Plants." *Science* 294 (2001) 2351–53.

Keller, Timothy J.. "Creation, Evolution and Christian Laypeople." BioLogos, November 1, 2009. http://biologos.org/uploads/projects/Keller_white_paper.pdf

———. *The Reason for God: Belief in an Age of Skepticism*. New York: Dutton, 2008.

Kent, George C., and Robert K. Carr. *Comparative Anatomy of the Vertebrates*. 9th ed. New York: McGraw Hill, 2001.

Kidner, Derek. *Genesis: An Introduction and Commentary*. Downers Grove, IL: InterVarsity, 1967.

Klicka, John, and Robert M. Zink. "The Importance of Recent Ice Ages in Speciation: A Failed Paradigm." *Science* 277 (1997) 1666–69.

Kline, Meredith. "Because It Had Not Rained." *Westminster Theological Journal* 20 (1958) 146–57

Klotz, John W. "Flora and Fauna of the Galapagos Islands." *Creation Research Society Quarterly* 9 (1972) 14–22.

Koonin, Eugene V. "The *Origin* at 150: Is A New Evolutionary Synthesis In Sight?" *Trends in Genetics* 25 (2009) 473–75.

Lamoureux, Denis O. *Evolutionary Creation: A Christian Approach to Evolution*. Eugene, OR: Wipf and Stock, 2008.

———. "Was Adam a Real Person? Part 1" BioLogos, September 2, 2010. http://biologos.org/blog/was-adam-a-real-person-part-i/

Larson, Edward J., and Larry Witham. "Scientists Are Still Keeping the Faith." *Nature* 386 (1997) 435–36.

Lewis, C.S. *Mere Christianity*. New York: Macmillan, 1955.

Liem, Karel, et al. *Functional Anatomy of the Vertebrates: An Evolutionary Perspective*. 3rd ed. Independence, KY: Cengage Learning, 2001.

Lioy, Daniel T. *Axis of Glory: A Biblical and Theological Analysis of the Temple Motif in Scripture*. New York: Peter Lang, 2010.

Livingstone, David N. *Darwin's Forgotten Defenders*. Grand Rapids: Eerdmans, 1987.

Lou, Zhexi. "In Search of the Whale's Sister." *Nature* 404 (2000) 235–37.

Louis, Ard. "Miracles and Science: The Long Shadow of David Hume." BioLogos, 2007. http://biologos.org/uploads/projects/louis_scholarly_essay.pdf

Machuga, Ric. *Life, the Universe and Everything: An Aristotelian Philosophy for a Scientific Age*. Eugene, OR: Wipf and Stock, 2011.

Matson, Dave E. "How Good Are Those Young-earth Arguments?" TalkOrigins Archive, December 10, 2002. http://www.talkorigins.org/faqs/hovind/howgood.html

Matzke, Nick. "Critique: 'Of Pandas and People.'" National Center for Science Education, November 23, 2004. http://ncse.com/creationism/analysis/critique-pandas-people

———. "Icon of Obfuscation." TalkOrigins Archive, January 23, 2004. http://www.talkorigins.org/faqs/wells/iconob.html

Max, Edward E. "Homology." Icons of Anti-Evolution, 2002. http://www.nmsr.org/text.htm#homo

———. "Plagiarized Errors and Molecular Genetics." TalkOrigins Archive, May 5, 2003. http://www.talkorigins.org/faqs/molgen/

———. "A Response to 'AFDave.'" TalkOrigins Archive, December 14, 2007. http://www.talkorigins.org/faqs/molgen/afdave.html

McGrath, Alister. *Surprised By Meaning: Science, Faith and How We Make Sense of Things.* Louisville: Westminster John Knox, 2011.

———. *Why God Won't Go Away: Is the New Atheism Running on Empty?* Nashville: Thomas Nelson, 2011.

McGrath, Alister, and Joanna C. McGrath. *The Dawkins Delusion?: Atheist Fundamentalism and the Denial of the Divine.* Downers Grove, IL: InterVarsity, 2010.

McIntyre, John A. "The Historical Adam." *Perspectives on Science and Christian Faith* 54 (2002) 150–57.

Meyer, Stephen C. *Darwin's Doubt: The Explosive Origin of Animal Life and The Case for Intelligent Design.* New York: HarperOne, 2013.

———. *Signature in the Cell: DNA and the Evidence for Intelligent Design.* New York, HarperOne, 2009.

Meyer, Stephen C., et al. *Explore Evolution: The Arguments For and Against Neo-Darwinism.* Melbourne: Hill House, 2007.

Michael J. Murray, and Jeffrey P. Schloss. "Evolution, Design, and Genomic Suboptimality: Does Science "Save Theology"?" *Proceedings of the National Academy of Science* 107 (2010) Vol 30, E121.

Miller, Kenneth R. *Finding Darwin's God: A Scientist's Search for Common Ground Between God and Evolution.* New York: Cliff Street, 1999.

———. "The Flagellum Unspun: The Collapse of "Irreducible Complexity." In *Debating Design: From Darwin to DNA*, edited by William A. Dembski and Michael Ruse, 81-97. New York: Cambridge University Press, 2008.

———. *Only a Theory: Evolution and the Battle for America's Soul.* New York: Viking Adult, 2008.

Mohler, R. Albert. "Why Does the Universe Look So Old?" *Arts & Facts* 39:10 (2010) 4-7.

Molley, Gerald. *Geology and Revelation.* New York, Putnam and Sons, 1870.

Moritz, Craig, et al. "Evolutionary Relationships Within the *Ensatina eschscholtzii* Complex Confirm the Ring Species Interpretation." *Systematic Biology* 41 (1992) 273–91.

Morris, Henry M. "Evolution is Religion — Not Science." *Arts and Facts* 30:2 (2001).

Morris, John D. "The Paluxy River Mystery." *Acts & Facts*, 15:1 (1986).

———. "Is There A Need For Creationist Research?" *Arts & Facts* 28:1 (1999).

Morrow, Jonathan. "How Should Christians Think About Evolution?" Think Christianly.org, November 22, 2010. http://www.youtube.com/watch?v=NAfAIY3ll4A

Muller, Gerd B., and Stuart A. Newman, editors. "Special Issue: Evolutionary Innovation and Morphological Novelty." *Journal of Experimental Zoology, Part B: Molecular and Developmental Evolution* 304B (2005) 485–631.

Murphy, *George L.* "Roads to Paradise and Perdition: Christ, Evolution and Original Sin." *Perspectives on Science and Christian Faith.* 58 (2006) 109-18.

Murphy, Nancey. "Divine Action In the Natural Order: Buridan's Ass and Schrodinger's Cat." In *Philosophy, Science and Divine Action*, edited by F. LeRon Shults et al., 263–304. Boston: Brill, 2009.

Murphy, Nancy, and Jeffrey Schloss. "Biology and Religion." In *Oxford Handbook of the Philosophy of Biology*, edited by Michael Ruse, 545–69. Oxford: Oxford University Press, 2008.

Murray, Michael. *Nature Red in Tooth and Claw: Theism and the Problem of Animal Suffering.* Oxford: Oxford University Press, 2011.

Musgrave, Ian. "Evolution of the Bacterial Flagellum." In *Why Intelligent Design Fails: A Scientific Critique of the New Creationism*, edited by Matt Young and Tanner Edis, 72–84. New Brunswick, NJ: Rutgers University Press, 2004.

National Academy of Sciences. *Science, Evolution and Creationism.* Washington, DC: National Academies Press, 2008.

National Association of Biology Teachers. *Statement on Teaching Evolution*, 1995. http://darwin.eeb.uconn.edu/Documents/NABT.html. The 2011 edition: http://www.nabt.org/websites/institution/index.php?p=92

National Center for Science Education. "Critique: Exploring 'Explore Evolution.'" October 17, 2008. http://ncse.com/creationism/analysis/explore-evolution

————. "The Sequence of Transitional Fossils." September 25, 2008. http://ncse.com/creationism/analysis/sequence-transitional-fossils

National Science Teachers Association. *The Teaching of Evolution.* Arlington: NSTA, 2003.

Nelson, Paul, and John M. Reynolds. "Young Earth Creationism." In *Three Views on Creation and Evolution*, edited by Moreland, James P. and John M. Reynolds, 41–75. Grand Rapids: Zondervan, 1999.

Nevins, Stuart. E. "Interpreting Earth History." *Impact*, March 1974.

Newcomb, Richard D., et al. 1997. "A Single Amino Acid Substitution Converts a Carboxylesterase to an Organophosporus Hydrolase and Confers Insecticide Resistance on a Blowfly." *Proceedings of the National Academy of Science* 94 (1997) 7464-68.

Newport, Frank. "In U.S., 46% Hold Creationist View of Human Origins." Gallup, June 1, 2012.

Niehaus, Jeffrey J. *Ancient Near East Themes in Biblical Theology.* Grand Rapids: Kregel Academic and Professional, 2008.

Nolan, David, et al. "HIV: Experiencing the Pressures of Modern Life." *Science* 307 (2005) 1422-24.

Noll, Mark A. "The Scandal of The Evangelical Mind." *Christianity Today*, October 25, 1993, 28-32.

Noll, Mark A., and David Livingston. "Charles Hodge and B.B. Warfield on Science, The Bible, Evolution and Darwinism." In *Perspectives on an Evolving Creation*, edited by Keith B. Miller, 61–71. Grand Rapids: Eerdmans, 2003.

————, eds. *B.B. Warfield: Evolution, Science and Scripture.* Grand Rapids: Baker, 2000.

Norell, Mark A., and Mick Ellison. *Unearthing the Dragons: The Great Feathered Dinosaur Discoveries*. New York: Pi, 2005.

Numbers, Ronald L. *The Creationists: The Evolution of Scientific Creationism*. Berkeley: University of California Press, 1992.

Ohlson, Kristin. "The End of Morality." *Discover*, July/August 2011.

Ohta, Yuriko, and Morimitsu Nishikimi. "Random Nucleotide Substitutions in Primate Nonfunctional Gene for L-Gulono-Gamma-Lactone Oxidase, the Missing Enzyme in L-Ascorbic Acid Biosynthesis" *Biochimica Biophysica Acta* 1472 (1999) 408–11.

Olofsson, Peter. "Intelligent Design and Mathematical Statistics: A Troubled Alliance." *Biology and Philosophy* 23 (2008) 543–53.

Origen. *De Principiis*. Translated from the Latin of Rufinus. http://www.ccel.org/ccel/schaff/anf04.vi.v.v.i.html

Oswalt, John N. *The Bible Among the Myths: Unique Revelation or Just Ancient Literature?* Grand Rapids: Zondervan, 2009.

Padian, Kevin, and Luis M. Chiappe. "The Origin of Birds and Their Flight." *Scientific American*, February 1998, 38–47.

Paley, William. *Natural Theology*. New York: American Tract Society, 1802.

Parkinson, William. "Questioning 'Flood Geology'" *Reports of the National Center for Science Education* 24:1 (2004).

Pelikan, Jaroslav, editor. *Luther's Works*,Vol. 1. St. Louis: Concordia, 1958.

Pennock, Robert. T. *Tower of Babel: The Evidence Against the New Creationism*. Cambridge, MA: MIT Press, 1999.

Peters, Ted, and Martinez Hewlett. *Can You Believe in God and Evolution? A Guide for the Perplexed*. Nashville: Abingdon, 2006.

Pew Research Center. "Religion A Strength and Weakness for Both Parties: Public Divided on Origins of Life." (Aug 30, 2005). http://people-press.org/report/254/religion-a-strength-and-weakness-for-both-parties

Pigliucci, Massimo, and Gerd B. Muller, editors. *Evolution: The Extended Synthesis*. Cambridge, MA: MIT Press, 2010.

Pinker, Steven. "Less Faith, More Reason." *The Harvard Crimson*, October 27, 2006.

Polkinghorne, John. *Science and Providence: God's Interaction with the World*. West Conshohocken, PA: Templeton Foundation, 2005.

———. *Science and Theology – An Introduction*. Minneapolis: Fortress, 1998.

Price, George M. *The New Geology*. Mountain View, CA: Pacific, 1923.

Prijambada, Irfan D., et al. "Emergence of Nylon Oligomer Degradation Enzymes in Pseudomonas aeruginosa PAO Through Experimental Evolution." *Applied and Environmental Microbiology* 61 (1995) 2020–22.

Prothero, Donald. R. *Evolution: What the Fossils Say and Why It Matters*. New York: Columbia University Press, 2007.

———. "Stephen Meyer's Fumbling Bumbling Cambrian Amateur Follies." 2013. http://www.amazon.com/review/R2HNOHERF138DU

Prum, Richard O., and Alan H. Brush. "Which Came First, The Feather Or The Bird?" *Scientific American*, March 2003, 84–93.

Purdom, Georgia, and Bodie Hodge. "Zonkeys, Ligers, and Wolphins, Oh My!" Answers in Genesis, August 6, 2008. http://www.answersingenesis.org/articles/aid/v3/n1/zonkeys-ligers-wholphins

Purves, William K., et al. *Life, the Science of Biology*, 7th ed. New York: W.H. Freeman, 2004.

Ramaswamy, Srinivas V., et al. "Single Nucleotide Polymorphisms in Genes Associated with Isoniazid Resistance in Mycobacterium tuberculosis." *Antimicrobial Agents and Chemotherapy* 47 (2003) 1241–50.

Ramm, Bernard. *The Christian View of Science and Scripture*. Grand Rapids: Eerdmans, 1954.

Rando, Oliver J., and Kevin J. Verstrepen. "Timescales of Genetic and Epigenetic Inheritance." *Cell* 128 (2007) 655–68.

Richards, Jay W. "Randy Isaac on 'Evolutionism.'" Evolution News and Views, October 26, 2010. http://www.evolutionnews.org/2010/10/randy_isaac_on_evolutionism039701.html

Ross, Hugh. *Creation and Time: A Biblical and Scientific Perspective on the Creation-Date Controversy*. Colorado Springs: NavPress, 1994.

———. *The Genesis Question: Scientific Advances and the Accuracy of Genesis*. 2nd ed. Colorado Springs, NavPress, 2001.

Rusbult, Craig. "Worldview Education in Public Schools." American Scientific Affiliation, 2003. http://www.asa3.org/ASA/education/views/balance.htm#i

Russell, Robert J. "Does 'the God Who Acts' Really Act? New Approaches to Divine Action in Light of Contemporary Science." In *Cosmology From Alpha to Omega: Towards the Mutually Creative Interaction of Theology and Science*, 130–77. Minneapolis: Fortress, 2008.

———. "Special Providence and Genetic Mutation: A New Defense of Theistic Evolution." In *Perspectives On An Evolving Creation*, edited by Keith B. Miller, 335–69. Grand Rapids: Eerdmans, 2003.

Sanchez-Villagra, Marcelo R., et al. "The Anatomy Of the World's Largest Extinct Rodent." *Science* 301 (2003) 1708–10

Sarfati, Jonathan. "How Did All the Animals Fit On Noah's Ark?" *Creation* 19:1 (1997) 16–19.

Schadewald, Robert. "The 1998 International Conference on Creationism." *Reports of the National Center for Science Education* 18 (1998) 22-25, 33.

Schloss, Jeffrey. "Evolution, Creation and the Sting of Death: A Reponse to John Laing." BioLogos, August 12, 2012. http://biologos.org/blog/series/southern-baptist-voices-evolution-and-death-series

———. "The Expelled Controversy: Overcoming or Raising Walls of Division?" American Scientific Affiliation, May 10, 2008. http://www.asa3.org/ASA/resources/Schloss200805.html

Schneider, Thomas D. "Dissecting Dembski's "Complex Specified Information." 2008. http://schneider.ncifcrf.gov/paper/ev/dembski/specified.complexity.html

Schroeder, Gerald L. *The Science of God: The Convergence of Scientific and Biblical Wisdom*. New York: Free Press, 1997.

Schwab, Ivan R. *Evolution's Witness: How Eyes Evolved*. Oxford: Oxford University Press, 2011.

Scientists in Congregations. A grant project "to catalyze the dialogue of theology and science in local congregation." http://www.scientistsincongregations.org/

Scott, Eugenie C. *Evolution vs. Creationism: An Introduction*. 2nd ed. Berkeley: University of California Press, 2009.

Senter, Phil. "The Defeat of Flood Geology by Flood Geology." *Reports of the National Center for Science Education* 31:3 (2011) 1–14.

———. "Vestigial Structures Exist Even Within the Creationist Paradigm." *Reports of the National Center for Science Education* 30:4 (2010).

Shallit Jeffrey, and Wesley R. Elsberry. "Playing Games with Probability: Dembski's Complex Specified Information." In *Why Intelligent Design Fails: A Scientific Critique of the New Creationism*, edited by Matt Young and Tanner Edis, 121–38. New Brunswick, NJ: Rutgers University Press, 2004.

Shanks, Niall, and Istvan Karsai. "Self-Organization and the Origin of Complexity." In *Why Intelligent Design Fails: A Scientific Critique of the New Creationism*. Edited by Matt Young and Tanner Edis, 85–106. New Brunswick, NJ: Rutgers University Press, 2004.

Shermer, Michael. *Why Darwin Matters: The Case Against Intelligent Design*. New York: Holt, 2007.

Sherwin, Frank. "Scientific Roadblocks to Whale Evolution." *Arts and Facts* 27:10 (1998).

Shimamura, Mitsuru, et al. "Molecular Evidence From Retroposons that Whales Form a Clade Within Even-Toed Ungulates." *Nature* 388 (1997) 666–70.

Shubin, Neil H., et al. "The Pectoral Fin of *Tiktaalik roseae* and the Origin of the Tetrapod Limb." *Nature* 440 (2006) 764–71.

———. "Pelvic Girdle and Fin of *Tiktaalik roseae*." *Proceedings of the National Academy of Science* 111 (2014) 881–82.

Snelling, Andrew A. "Doesn't the Order of Fossils in the Rock Record Favor Long Ages?" *The New Answers Book 2*. Chapter 31. Green Forest, AK: Master, 2010.

Sommer, Ralf J. "The Future of Evo-Devo: Model Systems and Evolutionary Theory." *Nature Reviews Genetics* 10 (2009) 416–22.

"Spirituality in America." *Newsweek*, August 29/September 5, 2005, 48–49.

Spurgeon, Charles H. *An All-Round Ministry*. London: Banner of Truth, 1960.

Strobel, Lee. *The Case for Faith*, Grand Rapids: Zondervan, 2000.

Sutera, Raymond. "Origin of Whales and the Power of Independent Evidence." TalkOrigins Archive, August 10, 2001. http://www.talkorigins.org/features/whales/

TalkDesign.org. A website that examines the scientific claims of Intelligent Design. www.talkdesign.org

TalkOrigins Archive. "The Age of the Earth FAQS." http://www.talkorigins.org/origins/faqs-youngearth.html

———. "Index to Creationist Claims." Edited by Mark Isaak, 2006. http://www.talkorigins.org/indexcc/list.html

Taylor, Paul F. "How Did Animals Spread All Over the World from Where the Ark Landed?" *The New Answers Book*, Chapter 11. Green Forest, AK: Master Books, 2006.

Teaching About Religion.org. "Dos and Don'ts." (2004). http://www.teachingaboutreligion.org/dosanddonts.html

The Ministry Theorum. "Engaging Science in the Life of Your Congregation." Calvin Theological Seminary. http://ministrytheorem.calvinseminary.edu/

Theobold, Douglas L. "29+ Evidences for Macroevolution: The Scientific Case for Common Descent." TalkOrigins Archive, last modified March 12, 2012. http://www.talkorigins.org/faqs/comdesc/

Thewissen, J.G.M., et al. "Fossil Evidence For the Origin of Aquatic Locomotion in Archaeocete Whales." *Science* 263 (1994) 210–12.

Thewissen, J.G.M., et al. "From Land to Water: the Origin of Whales, Dolphins and Porpoises." *Evolution: Education and Outreach* 2 (2009) 272–88.

———. "Skeletons of Terrestrial Cetaceans and the Relationship of Whales to Artiodactyls." *Nature* 413 (2001) 259–60.

Thompson, Timothy J. "Twenty-Four Young Earth Arguments Refuted." December 12, 2001. http://www.tim-thompson.com/young-earth2.html

van den Toren, Benno. "Not All Doctrines Are Equal—Configuring Adam and Eve." BioLogos, February 17, 2014. http://biologos.org/blog/not-all-doctrines-are-equalconfiguring-adam-and-eve

Venema, Dennis R. "Genesis and the Genome: Genomics Evidence for Human-Ape Common Ancestry and Ancestral Hominid Population Sizes." *Perspectives on Science and Christian Faith* 62 (2010) 166–78.

———. "Seeking a Signature." *Perspectives on Science and Christian Faith* 62 (2010) 276–83.

———. "A Tale of Three Creationists, Part 3." BioLogos, February 7, 2011. http://biologos.org/blog/a-tale-of-three-creationists-part-3

———. "Understanding Evolution: Is There "Junk" in Your Genome?" BioLogos, December 2011–February 2012. http://biologos.org/blog/series/understanding-evolution-is-there-junk-in-your-genome

———. "Understanding Evolution: Mitochondrial Eve, Y-Chromosome Adam, and Reasons to Believe." BioLogos, October 28, 2011. http://biologos.org/blog/understanding-evolution-mitochondrial-eve-y-chromosome-adam

Venema, Dennis, and Darrel Falk. "Signature in the Pseudogenes, Part 1." BioLogos, May 10, 2010. http://biologos.org/blog/signature-in-the-pseudogenes-part-1

Wake, David B., and Christopher J. Schneider. "Taxonomy of the Plethodontid Salamander Genus *Ensatina*." *Herpetologica* 54 (1998) 279–98.

Wallis, Claudia. "The Evolution Wars." *Time*, August 15, 2005.

Walton, John H. *Ancient Near Eastern Thought and the Old Testament: Introducing the Conceptual World of the Hebrew Bible*. Grand Rapids: Baker Academic, 2006.

———. *The Lost World of Genesis One: Ancient Cosmology and the Origins Debate*. Downers Grove, IL: InterVarsity, 2009.

Warfield, Benjamin B. "Calvin's Doctrine of Creation." In *The Works of Benjamin B. Warfield*, Vol 5. Oxford: Oxford University Press, 1931.

———. "Incarnate Truth." In *Selected Shorter Writings of Benjamin B. Warfield*, Vol 2, edited by John E. Meter. Philipsburg, NJ: Presbyterian and Reformed, 1970.

Weins, Roger C. "Radiometric Dating: A Christian Perspective." American Scientific Affiliation, 2002. http://www.asa3.org/ASA/resources/wiens.html

Weir, John T., and Dolph Schluter. "Ice Sheets Promote Speciation in Boreal Birds." *Proceedings of the Royal Society of London* 271 (2004) 1881–87.

Wells, Jonathan. *Icons of Evolution: Science or Myth?* Washington, DC: Regnery, 2000.

Westminster Confession of Faith. Lawrenceville, GA: Committee for Christian Education and Publications, 2007.

Whitcomb, John C., and Henry M. Morris. *The Genesis Flood*. Phillipsburg, NJ: Presbyterian and Reformed, 1961.

Whitcomb, John. C., and D.B. DeYoung. *The Moon: Its Creation, Form and Significance*. Winona Lake, IN: BMH, 1978.

Wimmer, Rainer, et al. "Direct Evidence for the Homo-Pan Clade." *Chromosome Research* 10 (2002) 55-61.

Wiseman, Jennifer. "Science as an Instrument of Worship." BioLogos, March 5–19, 2012. http://biologos.org/blog/series/science-as-an-instrument-of-worship

Witham, Larry. "Many Scientists See God's Hand in Evolution." *The Washington Times*, April 11, 1997, pA8.

Wood, Todd C. *Animal and Plant Baramins*. Eugene, OR: Wipf and Stock, 2008.

———. "The Current Status of Baraminology." *Creation Research Society Quarterly* 43:3 (2006) 149–58.

———. "The Truth About Evolution." September 30, 2009. http://toddcwood.blogspot.com/2009/09/truth-about-evolution.html

Young, Davis A. *The Biblical Flood — A Case Study of the Church's Response to Extrabiblical Evidence*. Grand Rapids, MI: Eerdmans, 1995.

Young, Matt, and Tanner Edis, editors. *Why Intelligent Design Fails: A Scientific Critique of the New Creationism*. New Brunswick, NJ: Rutgers University Press, 2004.

Zhang, Fucheng, et al. "Fossilized Melanosomes and the Colour of Cretaceous Dinosaurs and Birds." *Nature* 463 (2010) 1075-78.

Subject Index

Page numbers with "n" refer to footnotes, "f" refer to figures, and "t" refer to tables.

Scripture Index

Old Testament